United States
Nuclear Regulatory
Commission

United States
Army Corps
of Engineers

United States
Environmental
Protection Agency

NUREG/CP-0187
ERDC SR-04-2
EPA/600/R-04/117

Interagency Steering Committee on
Multimedia Environmental Models

I0482623

Proceedings of the

International Workshop on Uncertainty, Sensitivity, and Parameter Estimation for Multimedia Environmental Modeling

Proceedings of the

International Workshop on

Uncertainty, Sensitivity, and Parameter Estimation for Multimedia Environmental Modeling

Held August 19–21, 2003, at the U.S. Nuclear Regulatory Commission Headquarters
11545 Rockville Pike, Rockville, Maryland, USA

Under the sponsorship of the
Federal Working Group on Uncertainty and Parameter Estimation
of the
Federal Interagency Steering Committee on Multimedia Environmental Models
(ISCMEM)

Editors: Thomas J. Nicholson (NRC)
              Justin E. Babendreier (EPA)
              Philip D. Meyer (PNNL)
              Sitakanta Mohanty (CNWRA)
              Bruce B. Hicks (NOAA)
              George H. Leavesley (USGS)

DISCLAIMER

ABSTRACT

An *International Workshop on Uncertainty, Sensitivity, and Parameter Estimation for Multimedia Environmental Modeling* was held August 19–21, 2003, at the U.S. Nuclear Regulatory Commission Headquarters in Rockville, Maryland, USA. The workshop was organized and convened by the Federal *Working Group on Uncertainty and Parameter Estimation*, and sponsored by the Federal *Interagency Steering Committee on Multimedia Environmental Models* (ISCMEM). The workshop themes were parameter estimation, sensitivity analysis, and uncertainty analysis relevant to environmental modeling. The workshop objectives were to facilitate communication among U.S. Federal agencies conducting research on the workshop themes; obtain up-to-date information from invited technical experts; actively discuss the *state-of-the-science* in the workshop themes; and identify opportunities for pursuing new approaches. These objectives were met through the workshop presentations and discussions. The invited presenters focused on methods to identify, evaluate, and compare both existing and newly developed strategies and tools for parameter estimation, sensitivity and uncertainty analyses. Discussions explored how these strategies and tools could be used to better understand and characterize the sources of uncertainty in environmental modeling, and approaches to quantify them through comparative analysis of model simulations and monitoring. The presentations and discussions also focused on various approaches and applications of these strategies and tools, and specific lessons learned and research needs. In addition, the Memorandum of Understanding working group members and cooperators presented information and guidance for use in developing a common software application programming interface for methods and tools used in parameter estimation, sensitivity analysis, and uncertainty analysis.

CONTENTS

ACKNOWLEDGMENTS

The concept, planning, and execution of the international workshop and development of these proceedings were made possible through the vision and efforts of a group of volunteer members from the *Working Group on Uncertainty and Parameter Estimation*. Without their broad and continuing support of the workshop and its objectives, across the multiple Federal agencies represented, the workshop's scope, depth, and caliber of presentations simply would not have been possible. These Workshop Organizing Committee members were Justin E. Babendreier, Earl Edris, Bruce Hicks, George Leavesley, Philip Meyer, Sitakanta Mohanty, Thomas J. Nicholson, and Rien van Genuchten.

The Workshop Organizing Committee members wish to thank several key organizations and persons who helped to make the workshop the success it was. We would like to first acknowledge the invaluable support provided by the U.S. Nuclear Regulatory Commission (NRC) staff in workshop planning, correspondence, and facilitation. In particular, we wish to express our gratitude to the NRC for providing their auditorium and logistical support staff in Rockville, Maryland, to conduct the workshop. Lauren Claggett, Monique King, and Paula Garrity of the NRC staff were especially helpful in providing support prior to and following the workshop, which resulted in a well-managed meeting and the publication of these proceedings. Next, we wish to thank the NRC, U.S. Environmental Protection Agency, and U.S. Army Corps of Engineers, who provided the travel funding for the invited speakers. Finally, we are indebted to all of the 35 presenters, as listed in the table of contents, for their efforts, and the approximately 140 workshop attendees, many of whom are listed in Appendix D, who provided valuable feedback to the presenters during their presentations and discussion periods. Together, these presentations and discussions generated significant information and observations (as noted in the extended abstracts and session summaries documented in these proceedings), which made for a very successful workshop and report.

FOREWORD

These proceedings document presentations made at the *International Workshop on Uncertainty, Sensitivity, and Parameter Estimation for Multimedia Environmental Modeling* which was held August 19–21, 2003, at the U.S. Nuclear Regulatory Commission Headquarters in Rockville, Maryland, USA. The workshop was organized and convened by the Federal *Working Group on Uncertainty and Parameter Estimation* (WG2), and sponsored by the Federal *Interagency Steering Committee on Multimedia Environmental Models* (ISCMEM). ISCMEM was created through a Memorandum of Understanding (MOU) on cooperation and coordination of research, and development of multimedia environmental models, which includes the following eight Federal agencies:

- U.S. Nuclear Regulatory Commission (NRC), Office of Nuclear Regulatory Research (RES)
- U.S. Environmental Protection Agency (EPA), Office of Research and Development (ORD), National Exposure Research Laboratory
- U.S. Army Corps of Engineers (COE), Engineer Research and Development Center (ERDC)
- U.S. Department of Energy (DOE), Office of Science and Technology
- U.S. Department of the Interior (DOI), U.S. Geological Survey (USGS)
- U.S. Department of Agriculture (USDA), Agricultural Research Service (ARS)
- USDA, Natural Resources Conservation Service (NRCS)
- U.S. Department of Commerce (DOC), National Oceanic and Atmospheric Administration (NOAA)

As stated in the MOU, this initiative provides a mechanism for the cooperating Federal agencies to pursue a common technology in multimedia environmental modeling with a shared scientific basis. The MOU is intended to reduce redundancies and improve the common technology through exchange and comparisons of multimedia environmental models, software, and related databases. By entering into the MOU, the cooperating Federal agencies seek mutual benefit from their respective research and development programs related to multimedia environmental model development and enhancement activities, and ensure effective exchange of information between their technical staff and contractors. The International Workshop was organized by WG2 to help realize these goals.

The workshop themes were parameter estimation, sensitivity analysis, and uncertainty analysis relevant to environmental modeling. The workshop objectives were to facilitate communication among U.S. Federal agencies conducting research on the workshop themes; obtain up-to-date information from invited technical experts; actively discuss te state-of-the-science in the workshop themes; and identify opportunities for pursuing new approaches for parameter estimation, as well as sensitivity and uncertainty analyses. Theses proceedings summarize the workshop presentations as extended abstracts with accompanying information sources cited as selected references and Web sites. The workshop discussions were summarized by the WG2 members and cooperators and are documented in these proceedings. These proceedings completes the workshop objectives, and document the state-of-the-science in the workshop themes as presented by the U.S. Government scientists, contractors, and invited international experts.

These proceedings were reviewed and approved by the ISCMEM representatives of the eight participating Federal agencies under the MOU. The NRC published these proceedings with a NUREG/ CP document identifier. The document is also identified by EPA and COE-ERDC report numbers.

George Leavesley, Chair
Interagency Steering Committee on
Multimedia Environmental Modeling

1

INTRODUCTION AND OBJECTIVES

1. INTRODUCTION AND OBJECTIVES

Thomas Nicholson and George Leavesley

1.1 Background

The workshop was organized and convened by the Federal *Working Group (WG) on Uncertainty and Parameter Estimation* (WG2), and sponsored by the Federal *Interagency Steering Committee on Multimedia Environmental Models* (ISCMEM). The activities of the WG2 are defined by a Memorandum of Understanding (MOU) on research in multimedia environmental modeling. Specifically, the purpose of the MOU is to establish a framework to facilitate cooperation and coordination among the participating Federal agencies in research and development (R&D) of multimedia environmental models, software, and related databases, including development, enhancements, applications and assessments of site-specific, generic, and process-oriented multimedia environmental models as they pertain to human and environmental health risk assessment. The participating Federal agencies in the MOU are:

- U.S. Nuclear Regulatory Commission (NRC), Office of Nuclear Regulatory Research (RES)

- U.S. Environmental Protection Agency (EPA), Office of Research and Development (ORD), National Exposure Research Laboratory

- U.S. Army Corps of Engineers (COE), Engineer Research and Development Center (ERDC)

- U.S. Department of Energy (DOE), Office of Science and Technology

- U.S. Department of the Interior (DOI), U.S. Geological Survey (USGS)

- U.S. Department of Agriculture (USDA), Agricultural Research Service (ARS)

- USDA, Natural Resources Conservation Service (NRCS)

- U.S. Department of Commerce (DOC), National Oceanic and Atmospheric Administration

As stated in the MOU, this initiative provides a mechanism for the cooperating Federal agencies to pursue a common technology in multimedia environmental modeling with a shared scientific basis. The MOU is intended to reduce redundancies and improve the common technology through exchange and comparisons of multimedia environmental models, software, and related databases. By entering into the MOU, the cooperating Federal agencies seek mutual benefit from their respective R&D programs related to multimedia environmental model development and enhancement activities, and ensure effective exchange of information between their technical staff and contractors. These R&D programs include development and field applications of a wide variety of software modules, data processing tools, and uncertainty assessment approaches for understanding and predicting contaminant transport processes, including the impact of chemical and non-chemical stressors on human and ecological health.

The MOU focuses on exchange of information related to multimedia environmental modeling tools and supporting scientific information for environmental risk assessments, protocols for establishing linkages between disparate databases and models, and development and use of a common model-data framework. The MOU has facilitated the establishment of working partnerships among the technical staff and designated contractors of cooperating Federal agencies, in order to enhance productivity and mutual benefit through collaboration on mutually defined research studies such as the development of a common model-data framework. The goal of the MOU is to develop high-quality products using agreed-upon quality assurance (QA) and quality control (QC) procedures for environmental modeling.

The workshop conveners are members of the Federal *Working Group on Uncertainty and Parameter Estimation* (WG2). The objective of WG2 is to coordinate ongoing and new research that focuses on parameter estimation methods and uncertainty assessment strategies and techniques, in support of the development and application of environmental models. WG2 has the following goals:

- Develop a common understanding of the sources of uncertainty, and provide terminology.

- Identify, evaluate, and compare available uncertainty analysis strategies and tools.

- Develop new parameter estimation, sensitivity, and uncertainty analysis methodologies.

- Facilitate exchange of these techniques through technical workshops and professional meetings.

- Develop ways to better communicate uncertainty to decision-makers (e.g., visualization).

This workshop advanced the purpose and goals of the MOU and WG2. Some of the workshop presenters and attendees are members in another MOU working group on "Software System Design and Implementation," known as WG1. Other working groups under development focus on reactive transport modeling and watershed assessments.

A copy of the MOU can be viewed at the ISCMEM's public Web site, http://WWW.ISCMEM.Org. Specific details on the WG proposals, members, and activities; Steering Committee meeting minutes; and public meeting presentations and workshop proceedings are also available at the ISCMEM's Web site.

1.2 Workshop Objectives and Organization

In agreement with the WG 2 goals, this workshop was organized to (1) facilitate communication among U.S. Federal agencies conducting research on the workshop themes of parameter estimation, sensitivity analysis and uncertainty analysis; (2) obtain up-to-date information from invited technical experts; (3) actively discuss the state-of-the-science in the workshop themes; and (4) identify opportunities for pursuing new approaches for parameter estimation, as well as sensitivity and uncertainty analyses, related to multimedia environmental modeling.

The workshop was organized around the themes of parameter estimation, sensitivity analysis, and uncertainty analysis, with an emphasis on approaches, applications, lessons learned, and research needs. The workshop had five sessions that focused on these themes. International experts on these themes were identified and invited to make 30-minute presentations on their research. The session moderators and rapporteurs were WG2 members and cooperators. They prepared thematic introductions and discussion questions for each session, as reported in these proceedings. The table of contents follows the workshop agenda, and identifies the presenters and their presentations in their actual order. The concluding session focused on development of a common software application programming interface (API) for methods and tools used in parameter estimation, sensitivity analysis, and uncertainty analysis. This session was jointly developed and moderated by WG1 and WG2. Appendix A of these proceedings provides detailed guidance and information presented during this session.

1.3 Workshop Participation and Information Sources

Prior to and following the workshop, the WG2 members, speakers and workshop attendees identified many information sources including Web site links. Appendix B and Appendix C to these proceeding present these sources as a selected bibliography, and a listing of selected Web site links, respectively. Appendix D to these proceedings lists the workshop attendees by organization. Seven of the eight MOU participating Federal agencies were represented at the workshop. Each agency was given an opportunity to provide an overview of its specific needs and research on the workshop themes. In addition, Appendix E recounts the workshop agenda.

2

FEDERAL AGENCY OVERVIEWS OF PARAMETER ESTIMATION, SENSITIVITY, AND UNCERTAINTY APPROACHES

Overview of the NRC Research Program Related to Hydrologic Parameter Estimation, Sensitivity, and Uncertainty

Thomas J. Nicholson

Office of Nuclear Regulatory Research

U.S. Nuclear Regulatory Commission

Washington, D.C. 20555-0001

TJN@nrc,gov

The NRC's mission is to regulate the Nation's civilian use of by-product, source, and special nuclear materials to ensure adequate protection of public health and safety, promote the common defense and security, and protect the environment (NRC, 1997). One of the NRC's strategic performance goals is to ensure that its decisions are scientifically based, risk-informed, and shaped by operational experience, new information, and research, including cooperative international activities (NRC, 2000a). To help accomplish the NRC's mission and to support this strategic performance goal, the NRC maintains research capabilities to provide timely and independent technical bases for the agency's regulatory decisions. The NRC research objective dealing with radionuclide transport is to pursue more realistic and defensible estimates of exposure of the public to radiation from radionuclides released from contaminated sites or waste disposal facilities (NRC, 2002).

In the past, bounding estimates of the consequences of radionuclide transport from radioactive waste to humans were performed, but did not incorporate uncertainties and that made them difficult to defend (NRC, 2002). The NRC, in developing risk-informed, performance-based assessments, recognizes the need to address parameter and model uncertainties along with sensitivity analyses of the assumptions, processes, and parameters incorporated in the performance assessment models. Evolving approaches for estimating risk from releases of radioactive materials utilize computational tools that include parameter estimation, and sensitivity and uncertainty analyses. Uncertainty estimates are an important component in the decision-making process, and for communicating decisions to the public.

The NRC is funding research to develop a systematic approach for assessing hydrogeologic conceptual model and parameter uncertainties in multimedia environmental models (MEMs). The developed strategy is to identify and quantify uncertainties in alternative hydrogeologic conceptual models, parameter distributions, and assumptions in scenarios used in performance assessment models (NRC, 2000b). This research builds on accomplished methodologies developed by the University of Arizona on hydrogeologic conceptual model uncertainty (Neuman and Wierenga, 2003), and by Pacific Northwest National Laboratory on hydrologic and transport parameter uncertainty (Meyer and Gee, 1999; Meyer and Taira, 2001; and Meyer and Orr, 2002). The integrated approach will be tested using field datasets with sufficient information for comparing alternative conceptual/mathematical models and their attendant uncertainty.

The NRC is also funding research to examine the model abstraction process, and how complex and highly transient systems are represented. In particular, the study examines how abstraction techniques reduce the complexity of a simulation model while maintaining the validity of the simulation results with respect to the question that the simulation is being used to address (Frantz, 2003; and Pachepsky and others, 2003). Conventional ground-water flow and transport models simulate these complex systems through detailed numerical grids and associated data inputs, thereby introducing large computational and intensive data-collection requirements. Many of the detailed features, events, and processes represented in these complex models may have limited influence on the performance of a site. Model abstraction techniques could help identify those features, events, and processes that have a significant impact on site performance (Pachepsky and others, 2003). As such, they are useful to convey the level of conceptualization of the site that is essential for communication to both technical and lay audiences. Model abstraction techniques that can simplify and expedite the assessment of complex systems without significant loss of accuracy would greatly benefit the synthesis and review of performance assessments (Pachepsky and others, 2003). Thus, model abstraction reduces the complexity of a natural system to be simulated to its essential components and processes through a series of conceptualizations, selection of significant processes, and identification of the associated parameters.

Finally, the NRC is actively participating in the Memorandum of Understanding (MOU) on research into multimedia environmental modeling (ISCMEM, 2003). Through this participation, NRC staff and contractors are obtaining valuable information and tools. An important technical issue facing the application of MEMs is the inherent uncertainty associated with their conceptual/mathematical models and their parameter input estimates. Since many MEM applications involve an assessment of risk to the public health and/or environment, the use of uncertainty analysis techniques coupled to more robust parameter estimation methods would greatly enhance the insights and predictions derived from these models. Decisions involved in selecting and applying these uncertainty methods support the need for: (1) an a priori strategy which would systematically identify the various sources of uncertainty [e.g., lack of knowledge, natural variability, measurement or sampling error, randomness in "real-time" processes (Kundzewicz, 1995)]; and (2) an *a posteriori* strategy for comparing relative uncertainty estimates (e.g., conditional uncertainty measures or ranking of uncertainties). Many of the MOU participating agencies, notably the ARS, DOE, EPA, NOAA, NRC, and the USGS, are currently funding research studies related to this topic. Individually, these agencies also fund field studies, modeling assessments, and training courses related to MEMs.

New research that takes advantage of the MOU working group's activities and shared knowledge would facilitate development of a common understanding and technical framework to address the issues of uncertainty and parameter estimation. Therefore, the establishment of a unified methodology for addressing hydrologic conceptual, parameter, and scenario uncertainty is desirable. This methodology would build on and contribute to the MOU cooperative activities.

References:

Frantz, Frederick K., "A Taxonomy of Model Abstraction Techniques," Computer Sciences Corporation, One MONY Plaza, Mail Drop 37-2, Syracuse, NY, June 2003. (Available at the U.S. Air Force Research Laboratory's Web site: http://www.rl.af.mil/tech/papers/ModSim/ModAb.html)

Kundzewicz, Z.W., "Hydrological Uncertainty in Perspective," *in* Kundzewicz, Z.W. (ed), *New Uncertainty Concepts in Hydrology and Water Resources,* International Association of Hydrological Sciences—Proceedings of the International Workshop on New Uncertainty Concepts in Hydrology and Water Resources, held September 24–26, 1990, in Madralin, Poland, Cambridge University Press, New York, NY, 1995.

ISCMEM (Interagency Steering Committee on Multimedia Environmental Modeling), Public Web site: http://www.ISCMEM.Org , 2003. (MOU and details on the ISCMEM activities are provided.)

Meyer, P., and G. Gee, "Information on Hydrologic Conceptual Models, Parameters, Uncertainty Analysis, and Data Sources for Dose Assessment at Decommissioning Site," NUREG/CR-6656, U.S. Nuclear Regulatory Commission, Washington, DC, December 1999.

Meyer, P., and S. Orr, "Evaluation of Hydrologic Uncertainty Assessments for Decommissioning Sites Using Complex and Simplified Models," NUREG/CR-6767, U.S. Nuclear Regulatory Commission, Washington, DC, April 2002.

Meyer, P., and R.Y. Taira, "Hydrologic Uncertainty Assessment for Decommissioning Sites: Hypothetical Test Case Applications," NUREG/CR-6695, U.S. Nuclear Regulatory Commission, Washington, DC, February 2001.

Neuman, S.P., and P.J. Wierenga, "A Comprehensive Strategy of Hydrogeologic Modeling and Uncertainty Analysis for Nuclear Facilities and Sites," NUREG/CR-6805, U.S. Nuclear Regulatory Commission, Washington, DC, July 2003.

NRC, "Strategic Plan: Fiscal Year 1997 – Fiscal Year 2002," U.S. Nuclear Regulatory Commission, Washington, DC, September 1997.

NRC, "Strategic Plan: Appendix Fiscal Year 2000 – Fiscal Year 2005," NUREG-1614, Vol. 2, Part 2, U.S. Nuclear Regulatory Commission, Washington, DC, September 2000a.

NRC, "A Performance Assessment Methodology for Low-Level Radioactive Waste Disposal Facilities," NUREG-1573, U.S. Nuclear Regulatory Commission, Washington, DC, October 2000b.

NRC, "Radionuclide Transport in the Environment: Research Program Plan," Office of Nuclear Regulatory Research, U.S. Nuclear Regulatory Commission, Washington, DC, March 2002. (Available at NRC's Electronic Reading Room through Web-based access to ADAMS: http://www.nrc.gov/reading-rm/adams/web-based.html and search on ML020660731)

Pachepsky, Yakov, Martinus Th. van Genuchten, Ralph Cady and Thomas J. Nicholson, "Letter Report: Task 1 — Identification and Review of Model-Abstraction Techniques," U.S. Department of Agriculture, Agricultural Research Service, Beltsville, Maryland, February 27, 2003.

A Perspective from U.S. EPA: Uncertainty, Sensitivity, and Parameter Estimation In Multimedia Exposure and Risk Assessment Modeling

Justin E. Babendreier

Ecosystems Research Division, National Exposure Research Laboratory
Office of Research and Development, U.S. Environmental Protection Agency
Athens, Georgia 30605

babendreier.justin@epa.gov

Since its amalgamation as a Federal agency over 30 years ago, the U.S. Environmental Protection Agency (EPA) has undertaken many activities contributing to the international community's collective foundation for modern, multimedia environmental modeling. A key component of its current research agenda, the agency is seeking to better understand the role and functionality of multimedia modeling as an exposure/risk assessment tool to support sound decision-making.

Complimenting data collection, also a fundamental activity supporting its mission, EPA's complementary modeling efforts were initially focused on single-medium paradigms, which have formed, for the most part, the technical basis of many of today's regulatory programs. Over the last decade, EPA's assessment capabilities have matured into several integrated, multimedia-modeling software technologies that currently sit at or near deployment for use by both regulators and stakeholders. As these more complex, integrated assessment tools become engaged in the decision-making process, their use has underscored the need to more transparently characterize the attendant uncertainty in model inputs and outputs, and the associated sensitivity of model outputs to input error. Understanding, communicating, and optimally managing the strengths and weaknesses of integrated science, quantitatively captured as multimedia modeling technologies and data, is clearly one of the agency's most pressing challenges.

Discussions presented here on EPA's research perspectives for multimedia environmental modeling focus on several themes:

- Modern Environmental Assessments

- Probabilistic Exposure/Risk Assessments

- OMB-Driven Information Quality Guidelines

- Example Research Activities Being Conducted at USEPA/ORD/NERL/ERD

Multimedia Environmental Modeling

From a perspective of technology, multimedia modeling invokes the concept of a "modeling system," since the integration of many distinct, often single-medium, models is typically involved. These modeling systems include feed-forward only approaches that link black box models and data together. They also include more complicated modeling framework structures that more fully support "on-the-fly" feedback constructions between modeling components within these systems. In this genesis, EPA's research activities have spanned numerous technical areas including:

1. Research in core and applied science/engineering underpinning environmental models,

2. Data collection, estimation, and analysis,

3. Theoretical model development,

4. Model and modeling system technology development (i.e., software creation),

5. Computational systems R&D (e.g., high-end "mainframe" platforms, clusters, cyber-infrastructures)

6. Evaluation of modeling technologies (i.e., UA/SA/PE, quality assurance), and

7. Learning how best to communicate information to stakeholders and decision-makers.

Activities continue to be undertaken today by EPA to ultimately achieve integration of science-based modeling efforts, and to better inform evolving agency policy. The Office of Research and Development has engaged this overall approach as a means to best support regulatory-based decision-making, and achievement of EPA's overall mission to protect human health and the environment.

Modern Environmental Assessments

Representing the transition into modern environmental modeling assessments conducted today by the agency, and their use to support decision-making, increasingly one is found simultaneously deliberating upon:

- All potentially relevant media,

- All potentially relevant exposure pathways,

- Both human and ecological receptors,

- Variability, uncertainty, and sensitivity, and, overall,

- Validity, trustworthiness, and relevancy of our model predictions.

For the last two categories, which capture elements of model evaluation, we are beginning to view these as requisite steps in delivering quality assurance in model/system design for specific applications (Beck et al., 1997). Identifying, describing, and communicating uncertainty in an increasingly risk-based, model-driven, decision-making paradigm will continue to present a great challenge for the agency to meet over the coming decade.

A Multiplicity of Concerns in Decision-Making

As a further extension, in land-based waste management, for example, the agency's long-term research goal is currently formed upon the notion of developing easily deployed, integrated, science-based multimedia modeling technologies and data. These technologies will need ultimately to be able to address a multiplicity of concerns that manifest in decision-making, involving:

- Multiple media,

- Multiple pathways,

- Multiple receptors,

- Multiple pollutants, and

- Multiple scales (both spatial and temporal),

Site-based exposure and risk assessment is a common theme for much of the associated research being conducted. In view of the many model-based, site-based, decision-oriented problems faced today, there is first recognized an immediate priority for site-specific applications and demonstrations of existing multimedia decision-making technologies. In the near-term, key science enhancements and improved quality assurance in decision-making will also be moved forward.

Finally, there remains a need to further expand the capabilities of the existing base of multimedia decision-making technologies to more easily handle multi-scale and multi-pollutant constructions for model-based exposure and risk assessments. This will be particularly complex, for example, in attempting truly integrated risk assessment across multiple pollutants, since minimal data is available to guide the treatment of synergistic effects resulting from concurrent exposures. Efforts underway in Computational Toxicology research at EPA/ORD hold promise for expediting development of the information needed to bring such capabilities to a reliable point of functionality. There are, of course, many remaining, single-pollutant problems with pressing needs for improved science and data.

Probabilistic Assessments

Representing a slowly manifesting paradigm shift in agency approaches to modeling over the last 20 years, probabilistic-based exposure and risk assessments are today accepted by agency policy, and are increasingly common. The concept of integrating multimedia modeling and probabilistic assessment is also slowly making headway into model-supported decision-making. Objectivity, communication, familiarity, and decision-maker involvement are key issues that lay ahead. In general, better modeling hardware and software infrastructures are needed to conduct UA/SA/PE on a widespread scale, within and outside the agency, and to more easily interchange science and data across institutional boundaries.

OMB-Driven Information Quality Guidelines and CREM

In formulating regulation, the agency is increasingly held accountable today to formally demonstrate in its use of "influential information" (a) the assumptions used in an assessment; and (b) that the underlying science and data used are, to the extent practical, accurate, reliable, unbiased, and reproducible. This forms a basic tenant today of EPA's current Information Quality Guidelines (EPA, 2002), whose creation was itself guided by initiatives originating from the U.S. Office of Management and Budget (OMB).

There is added guidance on the subject of interacting with the public in matters relating to model evaluation tasks such as uncertainty analysis, sensitivity analysis, and parameter estimation (UA/SA/PE). As part of the 2002 guidelines, regulators must also establish that the presentation of information available is sufficiently comprehensive, informative, and understandable so as to allow the public to understand the risk assessment methodology and populations being considered, and the agency's plans for identifying and evaluating the uncertainty in risks. Specifically, allowing the public to determine:

- What populations are considered,

- How risk will be estimated for each,

- How uncertainty in risk will be quantified,

- Sources of uncertainty and ways to reduce it, and

- If peer-reviews support, are directly relevant to, or fail to support the assessment approach.

EPA Council on Regulatory Environmental Modeling

An example of a key agencywide effort underway, U.S. EPA's Council for Regulatory Environmental Modeling (CREM) is one of several supporting the implementation of EPA's Information Quality Guidelines. Comprised of representatives from across the agency, CREM is actively (EPA, 2003):

(1) Developing guidance for the development, assessment, and use of environmental models, and

(2) Collaborating with the U.S. National Academy of Sciences to develop recommendations for using environmental and human health models for decision-making.

UA/SA/PE Research Activities NERL/ERD

UA/SA/PE research being conducted at the Ecosystems Research Division of the National Exposure Research Laboratory is currently focused on the evaluation and development of innovative methods and associated software tools for conducting uncertainty and sensitivity analyses for simple and complex environmental models. Spanning theoretical and applied perspectives, this includes investigation of screening, local, and global analysis methods; parameter estimation techniques; model calibration strategies; statistical sampling methods; and parameter distribution transformations.

Algorithms are evaluated in the context of performing single-medium and multimedia fate and transport modeling, typically coupled with model-based exposure and risk assessments addressing ecological and human health concerns. Techniques that show promise in advancing the ability to quantify uncertainty and sensitivity for low and/or high order environmental models receive additional focus in learning how the methods might best be implemented within supportive modeling frameworks. To facilitate model simulation experimentation in this research program, a 180+ node PC-based, Windows/Linux-based supercomputing hardware and software infrastructure was also developed.

Summarizing the specific focuses in NERL/ERD's research program, activities include:

- Uncertainty Analysis

  o Sampling-based: Integrated High-Order Models

  o N-Dimensional Iterator (e.g., 2-stage MC)

  o Model Error and Modeler Error Quantification

- Sensitivity Analysis and Parameter Estimation

  o Screening-Level (Andres' IFFD, Morris's Oat)

  o Local (JUPITER; Integration of Inverse Problem Technologies)

  o Global (Correlation/Regression, RSA, TSDE)

  o SA-based Performance Validation

- PC-based Windows/Linux Supercomputing for UA/SA/PE.

Beck, M.B., Ravetz, J.R., Mulkey, L.A., Barnwell, T.O.. (1997). On the Problem of Model Validation for Predictive Exposure Assessments. Stochastic Hydrology and Hydraulics, 11:229-254.

EPA (2002). Guidelines for Ensuring and Maximizing the Quality, Objectivity, Utility, and Integrity of Information Disseminated by the Environmental Protection Agency. Office of Environmental Information. EPA/260R-02-008, http://www.epa.gov/quality/informationguidelines/index.html.

EPA (2003). Draft Guidance on the Development, Evaluation, and Application of Regulatory Environmental Models. Office of Research and Development, Office of Science Policy, Council for Regulatory Environmental Modeling (CREM), http://cfpub.epa.gov/crem/cremlib.cfm.

USGS Overview of Research Activities for Evaluating Uncertainty

Mary C. Hill, U.S. Geological Survey, Boulder, CO, USA
George Leavesley, U.S. Geological Survey, Lakewood, CO, USA

The USGS serves the Nation by providing reliable scientific information to
- describe and understand the Earth;
- minimize loss of life and property from natural disasters;
- manage water, biological, energy, and mineral resources; and
- enhance and protect our quality of life.

The USGS mission is accomplished by offices, personnel, and projects located in all 50 states and several foreign countries. Some projects are funded federally; others are supported in part, or in whole, by other governmental entities such as states, counties, cities, and foreign governments.

In pursuit of its mission, the USGS collects, manages, and analyzes a wide range of environmental data. Much of the data is displayed online. For example, real-time surface-water data are presented at HYPERLINK "http://water.usgs.gov/waterwatch/" http://water.usgs.gov/waterwatch/, and national maps of geology, hydrology, land-use, and biological resources are presented at HYPERLINK "http://nationalmap.usgs.gov" http://nationalmap.usgs.gov. Many societal decisions and scientific efforts rely on USGS databases.

The USGS develops a wide range of public-domain, open-source software, which can be accessed through HYPERLINK "http://www.usgs.gov/pubprod/software.html" http://www.usgs.gov/pubprod/software.html. In some fields, USGS software has become the standard. For example, the MODFLOW ground-water model (Harbaugh and others, 2000; Hill and others, 2000) is used widely throughout the US. In other countries, it has been used for as much as 90% of numerical ground-water studies.

When modeling environmental systems, quantifying uncertainty of simulated results requires detailed analysis at every step of system characterization, simulation, and calibration. This includes understanding and quantifying errors and variability in data collection and interpretation, conceptual model development, mathematical formulation, parameter estimation, and numerical calculation. Analysis of uncertainty has a long and enduring tradition in the USGS. For example, Carter and Anderson (1963) used repeated measurement of selected reaches to determine that errors of about 5% are typical of even good streamflow measurements. In computer modeling, Cooley (1977) was one of the first to consider the utility of regression-based methods to improve how data are used and uncertainty is accounted for in models of complex environmental systems. Thise effort has advanced through the development of methods and software for sensitivity analysis, data-needs assessment, calibration, and uncertainty evaluation related to many environmental systems [for example, Moss and Lins (1989), Leavesley and others (1996), Poeter and Hill (1998), Hill (1998), Parkhurst and Appelo (1999), Hill and others (2001), Helsel and Hirsch (2002), and Nordstrom (2004)].

The most recent effort is the JUPITER project (Joint Universal Parameter IdenTification and Evaluation of Reliability) being developed in collaboration with the US EPA. It owes its existence, in part, to collaborations encouraged by ISCMEM. JUPITER is composed of an API (Application Programming Interface) from which application programs are constructed. It is designed to encourage contributions from many scientists, and for these methods to be readily available to all modelers. In this way, alternative methods can be readily compared in the context of practical problems. Such comparisons will facilitate further developments and evaluation and, thereby, provide useful, proven and timely approaches to resource managers. One initial JUPITER application, J_MMRI, includes methods for multi-model ranking and inference, and was used to test the AICc method [Poeter and Anderson (2004)].

The USGS believes strongly that the nation's environmental problems can be addressed most effectively through cooperation between federal agencies such as that encouraged by ISCMEM, and intends to continue its participation in what has been a very fruitful endeavor.

References:

Anning, D. W., 2002, Standard errors of annual discharge and change in reservoir content data from selected stations in the lower Colorado River streamflow-gaging station network 1995-99: U.S. Geological Survey Water-Resources Investigations Report 2001-4240, 81p. HYPERLINK "http://pubs.er.usgs.gov/pubs/wri/wri014240" http://pubs.er.usgs.gov/pubs/wri/wri014240

Carter, R.W. and Anderson, I.E., 1963, Accuracy of current meter measurements: American Society of Civil engineers Journal, v. 89, no. HV4, p. 105-115.

Harbaugh, A. W., Banta, E. R., Hill, M. C., and McDonald, M. G., 2000, MODFLOW-2000, The U.S. Geological Survey modular ground-water model – Users guide to modularization concepts and the ground-water flow process: U.S. Geological Survey Open-File Report 00-92, 121 p. HYPERLINK "http://water.usgs.gov/ nrp/gwsoftware/modflow2000/modflow2000.html" http://water.usgs.gov/nrp/gwsoftware/modflow2000/ modflow2000 html

Helsel, D.R., and Hirsch, R.M., 2002, Statistical methods in water resources: U.S. Geological Survey Techniques in Water Resources, Book 4, Chapter A3, 510 p, HYPERLINK "http://pubs.water.usgs.gov/twri4a3" http://pubs. water.usgs.gov/twri4a3 .

Hill, M. C., 1998, Methods and guidelines for effective model calibration: U.S. Geological Survey Water-Resources Investigations Report 98-4005, 90 p. HYPERLINK "http://pubs.water.usgs.gov/wri984005/" http://pubs.water. usgs.gov/wri984005/

Hill, M. C., Banta, E. R., Harbaugh, A. W., and Anderman, E. R., 2000, MODFLOW 2000, The U.S. Geological Survey modular ground-water model, User's guide to the observation, sensitivity, and parameter-estimation processes: U.S. Geological Survey Open-File Report 00-184, 209p. HYPERLINK "http://water.usgs.gov/ nrp/gwsoftware/modflow2000/modflow2000.html" http://water.usgs.gov/nrp/gwsoftware/modflow2000/ modflow2000 html

Hill, M.C.; Ely, D.M.; Tiedeman, C. R.; O'Brien, G.M.; D'Agnese, F.A.; Faunt, C.C., 2001,Preliminary evaluation of the importance of existing hydraulic-head observation locations to advective-transport predictions, Death Valley regional flow system, California and Nevada: U.S. Geological Survey Water-Resources Investigations Report 2000-4282, 82p. HYPERLINK "http://water.usgs.gov/pubs/wri/wri004282/" http://water.usgs.gov/pubs/ wri/wri004282/

Leavesley, G.H., Restrepo, P.J., Markstron, S.L., Dixon, M., and Stannard, L.G., 1996, The Modular Modeling System (MMS), User's Manual: U.S. Geological Survey Open-File Report 96-151, 142p. HYPERLINK "http:// pubs.er.usgs.gov/pubs/ofr/ofr96151" http://pubs.er.usgs.gov/pubs/ofr/ofr96151

Moss, Marshall E.; Lins, Harry F., 1989, Water resources in the twenty-first century; a study of the implications of climate uncertainty: U.S. Geological Survey Circular 1030, 25p.

Nordstrom, D.K., 2004, Modeling low-temperature geochemical processes, in Treatise on geochemistry, H.D. Holland and K.K.Turekian, ex. eds.: vol. 5, Surface and ground water, weathering, and soils, J.I. Drever, ed., Elsevier Pergamon, Amsterdam, p. 37-72.

Parkhurst, D.L. and Appelo, C.A.J., 1999, User's guide to PHREEQC (Version 2) : a computer program for speciation, batch-reaction, one-dimensional transport, and inverse geochemical calculations: U.S. Geological Survey Water-Resources Investigations Report 99-4259, 312p. HYPERLINK "http://wwwbrr.cr.usgs.gov/ projects/GWC_coupled/phreeqc/index.html" http://wwwbrr.cr.usgs.gov/projects/GWC_coupled/phreeqc/index. html

Poeter, E.P. and Hill, M.C., 1998, Documentation of UCODE, A computer code for universal inverse modeling: U.S. Geological Survey Water-Resources Investigations Report 98-4080. 116p. HYPERLINK "http://water.usgs.gov software/ucode html" http://water.usgs.gov/software/ucode.html

Poeter, E.P. and Anderson, D.R., 2004, Multi-model ranking and inference in groundwater modelling, in Kovar, Karel and Hrkal, Z., eds, Finite-Element Models, MODFLOW, and More, Solving groundwater problems, Proceedings, Carlsbad, Czech Republic, September 13-16, 2004, p. 85-89.

NOAA Overview:

Uncertainty in Multimedia Modeling Applications

Bruce B. Hicks

Air Resources Laboratory
Office of Oceanic and Atmospheric Research, NOAA
1315 East West Highway, Silver Spring, Maryland 20910

The principal mission of NOAA relates to the provision of environmental forecasts, with emphasis on the atmosphere and the hydrosphere. Diminishing water resources and the susceptibility to flooding elevate the accurate prediction of water availability to a critical level. Rainfall and snowmelt are often key considerations. In practice, precipitation at a single location is one of the most variable phenomena of nature, and prediction of it is necessarily probabilistic. As the averaging area increases, uncertainty decreases, but it remains that forecasting of floods must have a strong probabilistic component. The predictive models on which water availability and flood forecasting rely must take all of the related uncertainties into account and propagate them accurately through the overall environmental system. Add the uncertainties of snowmelt to the mix and we finish up with a highly complex mix of deterministic and stochastic processes.

The chemical composition of the precipitation is of increasing interest, since recent assessments have shown that as much as 40% of the nutrient influx into coastal ecosystems might be due to deposition from the atmosphere after transport from pollution sources far upwind. The classical view of coastal ecosystem decline is being revised. No longer is the focus of regulatory efforts only on point sources with discharges into the water body in question, but it is also on the consequences of distant emissions that are transported to the catchment area through the atmosphere. Once again, the precipitation process is centrally involved (although we must also consider dry deposition, a slow but continual process whereas wet deposition by rain is far more efficient but highly intermittent). NOAA has elevated the forecasting of ecosystem health to a high priority, requiring a new focus on the way in which pollution from all sources affects sensitive areas, primarily along the coasts. The development of multimedia models is essential. Several target areas are being identified for initial attention, such as the Great Lakes, the Gulf of Mexico, and the mid-Atlantic coast. The focus is on both long-term "chronic" aspects of the problem and on short-term "acute" considerations. In the long-term case, the key product is likely to be the accurate prediction of trends with time. In the short-term case, the need may well be the prediction of the probability that damaging levels will be exceeded. In both contexts, uncertainties and natural variability must be taken into account. In all cases, the consequences to the living environment must be considered. The breadth of the research in NOAA stretches from the transport of pollutants from emission sources to the health of the fish in the estuaries that are eventually affected.

DOE Overview

Beth Moore, DOE

A presentation on the section heading topic was given by the speaker identified.
No abstract was provided.

The USDA-Agricultural Research Service Watershed Research Program

Mark A. Weltz and Dale A. Bucks

USDA-ARS, National Program Staff

5601 Sunnyside Ave., Beltsville, Maryland

Phone: (301) 504-4600 and Fax: (301) 504-6231

Abstract

Water quantity and quality issues have increasingly become the focus of attention of United States citizens, private and public organizations, and units of government striving to meet competing demands while protecting the environment and public health. Sound agricultural management practices are required to ensure success in maintaining a healthy and productive land and water base that sustains local communities, food and fiber production, and also protects and restores critical natural systems. The central mission of the USDA-Agricultural Research Service (ARS) Watershed Research Program is to address challenges and solve problems that confront American agriculture enterprises. The ARS accomplishes this mission by using the scientific method to improve our understanding of basic hydrologic processes. ARS and its collaborators use this knowledge to develop new methodologies and technologies to mitigate deleterious effects of floods and droughts, reduce soil erosion and sedimentation on our farms and within our streams and lakes, improve water quality, and enhance water supply and availability. The ARS watershed network is a set of geographically distributed experimental watersheds that has been operational for more than 70 years and is the most comprehensive watershed network of its kind in the world. The watershed facilities serve as outdoor laboratories that provide an essential research capacity for conducting basic long-term, high-risk field research. The watershed network and its associated historical database from 23 States provide the only means to evaluate the long-term impacts and benefits of implementing agricultural practices on water quality and water availability, documenting effects of global change, and developing new instrumentation and decision support systems to enhance the economic and environmental sustainability of agriculture. More than 140 ARS subwatersheds and related facilities, ranging in size from 0.2 hectares to over 600 km^2, are currently operated from 17 research facilities within the continental United States.

Introduction and History

The ARS Watershed Network (Figure 1) can be broadly characterized as an *intensive* network where some sets of geographically distributed watersheds are observed and studied in great detail. In an intensive network, numerous observations and dense instrumentation nets are concentrated in relatively small watersheds to support investigations for specific hydrologic process understanding. The ARS Experimental Watershed Program grew out of depression era efforts by the Civil Conservation Corps (CCC) and the Soil Conservation Service (SCS). Kelly and Glymph (1965) described the early history of the watershed program, including research associated with the 1930's conservation motto "stop the water where it falls." The research focused on merits of upstream watershed conservation to reduce runoff and erosion.

There was early recognition of the scaling problems in transferring knowledge from small to larger watersheds (Harrold and Stephens, 1965). This problem and growing concern of downstream, offsite impacts of upstream watershed practices resulted in establishment of a subset of larger ARS experimental watersheds associated with new watershed research centers in a number of hydroclimatic regions in compliance with U.S. Senate Document 59 (Great Plains, Northeast, Northwest, Southeast and Southwest Watershed Research Centers in Chickasha, OK; State College, PA; Boise, ID; Tifton, GA; and Tucson, AZ; respectively). The goal in establishing the watershed research centers was to select a representative basin and establish satellite basins, which were less well instrumented, to extend the data and findings from the primary watershed center. Nested watersheds and unit source areas on major soil types were included in the watershed designs to investigate scale effects.

The Current Network

Seventeen locations within the contiguous United States are currently collecting a variety of abiotic and biotic data at 140 subwatersheds nested within the larger ARS watersheds. The ARS watersheds represent numerous diverse land uses and agricultural practices and cover a wide range of hydroclimatic conditions. The diversity of observations made at these watersheds is a reflection of the diversity in dominant hydroclimatic processes across locations and evolving research objectives. As research objectives have changed to address problems such as water quality (e.g., biotic, chemical, pathogen, sediment) and global change, instrumentation and observations have been added to the basic rainfall-runoff observation infrastructure. An important component of the network is the ARS Hydraulics Engineering Unit located in Stillwater, Oklahoma, which has provided critical expertise and facilities in the development of flood-control and hydraulic structures and runoff measurement devices deployed in many of the watersheds. ARS also conducts hydraulics engineering research on the design and safety issues related to earthen dam flood control structures in support of Public Laws' PL-534 and PL-566 at Stillwater, Oklahoma.

Data Availability

The Agricultural Research Service (ARS) is a research organization. Data collected from the ARS Watershed Network should be considered *experimental data*. While much of the original instrumentation, installation, and data processing procedures for basic rainfall, runoff, and meteorological data was guided by Handbook 224 (Brakensiek et al., 1979), data collection has evolved at individual locations to address regional research needs. ARS watershed data have not historically been collected and reviewed under a national standard set of guidelines and procedures such as those employed by the USGS. Instruments, parameters observed, and data reduction procedures vary from watershed to watershed. A description of data acquisition programs and an assessment of the quality of collected data at many of the experimental watersheds is described in USDA (1982).

Based on data compiled and maintained by Jane Thurman at the Hydrology and Remote Sensing Laboratory in Beltsville, Maryland, as of January 1, 1991, ARS had operated over 600 watersheds in its history. A rainfall-runoff database is available from the Hydrology and Remote Sensing Laboratory for 333 of these watersheds. About 16,600 station-years of data are stored there from watersheds ranging from 0.2 hectares to 12,400 km^2. After 1990, the HRSL no longer archived data but has provided links back to the individual ARS watershed locations. These locations are making a concerted effort to make the ARS Experimental Watershed data more readily accessible and to provide additional types of data (soils, vegetation maps, geology in standard geographic information system formats, etc.) available through a Web-enabled search-and-retrieval system, but progress varies due to resource constraints. It is anticipated that a prototype system that is currently being developed will be available in late 2005.

Major Accomplishments

Development of innovative instrumentation: ARS watersheds have pioneered the testing and development of stream flow instrumentation including the drop-box weir for high-energy, high-bedload systems, supercritical flumes for arid regions, and small-scale runoff flumes. Stream sampling methods for water quality such as the Coshocton Wheel, traversing slot sediment samplers, and widely used in-stream samplers have also come from ARS watersheds. Other advances include state-of-the-art hydro-meteorological field sensors, watershed-wide telemetry, archival equipment and systems, the dual-gage precipitation measurement system, load cell precipitation gage, radar and acoustics technology to measure sediment transport, snow pillow and advanced snow sensors and programable, variable rate, rainfall simulators.

Development and testing of remote sensing technologies and applications: Pioneering research in both the theory and application of remote sensing to the use of microwave remote sensing of soil moisture has been conducted by ARS personnel at the ARS watersheds. Both NASA and the Japanese space agency are currently implementing results. Large-scale soil moisture observations may contribute to major breakthroughs for hydraulic modeling, crop yield forecasting, drought assessment, irrigation management and the ability to detect and model land surface response in climate change studies. In addition, long-term acquisition of complimentary remote sensing imagery supported by ground and atmospheric measurements at several ARS watersheds are used as long-term validation for both NASA and European Space Agency sensors.

Improvement in agricultural water quality: Nutrients and herbicides related to farming practices have been detected in shallow groundwater and agricultural runoff in many parts of the country. ARS watershed research has led to (i) buffer system designs composed of grasses and trees that can be used to assimilate nitrogen and phosphorus from both surface water and shallow groundwater and reduce offsite impacts of animal feeding operations, (ii) nitrogen management practices, using the ARS- developed Late Spring Nitrate Test, which have demonstrated reduced nitrate pollution levels, (iii) the development of the SWAT (Soil and Water Assessment Tool) model, which has been applied extensively for policy planning and in developing best management practice alternatives, and (iv) the quantification of water quality impacts of brush control herbicides picloram and clopyralid, which were shown to dissipate quickly in the soil and to be undetectable in surface runoff or subsurface flow. Studies in ARS watersheds were instrumental in obtaining approval of these herbicides for public use.

Rainfall frequency analyses: Analyses of ARS dense raingauge networks were utilized to modify NOAA National Atlases of rainfall frequency that is utilized to develop design storm characteristics for flood control maps and prevention activities.

Development of hydrologic and natural resource management models: ARS watershed research and data have been critical to the development and validation of natural resource models too numerous to mention in this report in detail (ANAGNPS, CONCEPTS, CREAMS, Curve Number, GLEAMS, EPIC, KINEROS, REMM, RUSLE2, SRM, SWAT, and WEPP). An example of an ARS model that has had tremendous impact is the USLE (Universal Soil Loss Equation) model. The USLE and its replacements the Revised Universal Soil Loss Equation (RUSLE) and RUSLE2 erosion prediction tools are the most widely utilized field scale erosion prediction tools in use around the world today. The American Society of Agricultural Engineering recently recognized the USLE model for its outstanding impact on sustaining agriculture production around the world by reducing soil loss. The ARS-developed KINEROS model was utilized by a consulting firm and resulted in construction savings of over $16 million on a series of dams on the Au Sable River in Michigan. More recently, the SWRRB (Simulator for Water Resources in Rural Basins) mode and the SWAT (Soil Water Assessment Tool) model have been used by many Federal and State agencies to evaluate USDA conservation program effectiveness and the economic and environmental impacts/benefits derived from implementing conservation practices.

Hydraulic Structure Design: The Natural Resources Conservation Service (NRCS) has used ARS developed procedures for design and construction of more than 800,00 km (500,000 mi) of vegetated channels. The American Society of Agricultural Engineering lists the design procedure as one of the top five outstanding agricultural engineering achievements of the 20th century. These and other design criteria are available on the SITES 2000: Water Resources Site Analysis CD from ARS. This expert system is helping NRCS and local sponsors of earthen dam flood control structures design urgently needed safety upgrades to the 11,000 structures that have been constructed across the United States. ARS in association with the Oklahoma Conservation Commission has also developed a video that describes the benefits of these small hydraulic structures that explains the importance of maintenance and repair of the structures.

The ARS Watershed Program and its Experimental Watersheds provide exceptional "**outdoor laboratories**" to develop knowledge that addresses societal water resource issues in real world settings. The stability of these research platforms, with a high-quality knowledge base and observational infrastructure makes them ideal facilities for collaborative research to investigate the hydrologic cycle and potential changes to it across a wide range of hydro-climatic conditions. There is no comparable network of experimental agricultural watersheds in the world.

References

Brakensiek, D.L., H.B. Osborn, and W.J. Rawls, coordinators. 1979. Field manual for research in agricultural hydrology. U.S. Dept. Agric., Agric. Handbook 224, 550 p.

Harrold, L.L., and J.C. Stephens. 1965. Experimental watershed for research on upsteram surface waters. IASH, Sym. of Budapest, Representative and Experimental Areas, IASH Pub. No. 66, Vol. 1, p. 39–53.

Kelly, L.L., and L.M. Glymph. 1965. Experimental watersheds and hydrologic research. IASH, Sym. of Budapest, Representative and Experimental Areas, IASH Pub. No. 66, Vol. 1, p. 5–11.

U.S. Dept. of Agriculture. 1982. The Quality of Agricultural Research Service Watershed and Plot Data. Ed. by C.W. Johnson, D.A. Farrell, and F.W. Blaisdell, Agric. Reviews and Manuals, ARM-W-31, 168 p.

This material was originally presented by Dave Goodrich, Daniel Marks, Mark Seyfried, and Clarence Richardson as a poster at the December 2000 America Geophysical Union in San Francisco, California and has been updated for this meeting. We would also like to thank Jane Thurman and all the other ARS employees in the watershed program for the work they have put in developing and maintaining the historical watershed data.

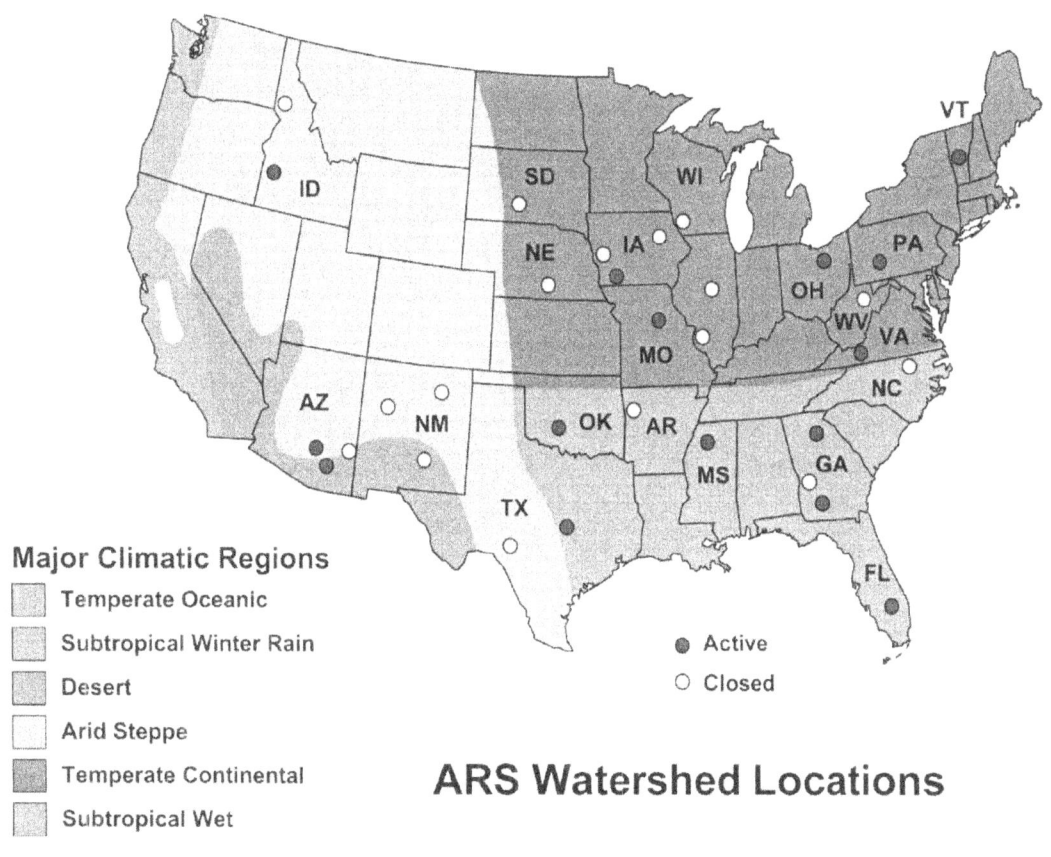

Major Climatic Regions

- Temperate Oceanic
- Subtropical Winter Rain
- Desert
- Arid Steppe
- Temperate Continental
- Subtropical Wet

● Active
○ Closed

ARS Watershed Locations

Figure 1. Locations of the historical and active Agricultural Research Service experimental watersheds.

USACOE Overview

Earl Edris, USACOE

A presentation on the section heading topic was given by the speaker identified. No abstract was provided.

3

SESSION 1:

PARAMETER ESTIMATION APPROACHES, APPLICATIONS, AND LESSONS LEARNED — *IDENTIFICATION OF RESEARCH NEEDS*

Overview and Summary

Editor: Philip Meyer

The first session of the workshop was comprised of eight presentations addressing parameter estimation methods and the interfaces between parameter estimation and sensitivity and uncertainty analyses. Two general types of parameter estimation, as distinguished by the methods used, were discussed. The first of these involves the application of optimization methods to determine parameters based on measurements of system response, that is, the quantities being simulated by a model (e.g., hydraulic head in a groundwater model, stream discharge in a surface water model, concentration in a transport model). In this category, two approaches have been adopted. The first approach integrates parameter estimation within the application model. Examples include the HYDRUS code, a model of variably saturated flow, solute transport, and heat movement in porous media, and the popular groundwater modeling code MODFLOW-2000. The second approach implements parameter estimation methods independently of the application model, for example by interacting with the model's input and output files as is done in the codes PEST and UCODE. The performance of the regression does not depend on whether the parameter estimation is integrated or not. A new application programming interface (the JUPITER API) being developed to support parameter estimation, sensitivity analysis, and uncertainty assessment was described. Applications using this API are being developed, including the next generation of UCODE. One of the difficulties with optimization-based parameter estimation is the problem of nonunique solutions. Methods to address this difficulty were discussed by a number of presenters and included incorporation of prior parameter information, regularization, and multi-criteria optimization.

The second general type of parameter estimation takes place when a model is parameterized without access to measurements of system response. Parameters can be estimated under these conditions by extrapolating from knowledge of parameters at other representative sites, by indirect estimation using relationships between parameter values and system characteristics that have been measured (or can be more easily measured than system response), and by direct measurement. The Prediction in Ungauged Basins (PUB) and Model Parameter Estimation Experiment (MOPEX) programs address parameter estimation in surface water (and atmospheric) models using primarily extrapolation and indirect parameter estimation methods. The Rosetta code can be used to estimate soil hydraulic parameters indirectly for variably saturated flow models using more easily measured quantities (soil texture, bulk density, and water content at specific pressures). Application of a surface complexation model for uranium adsorption was discussed. Estimation of the parameters for this model relied on extrapolation from a limited set of laboratory experiments. A description was provided of methods (generally extrapolation or indirect estimation) used by NRC staff to estimate parameters for dose analyses at decommissioning sites.

The application of sensitivity information in the modeling process was discussed by a number of presenters. This sensitivity information may arise from the parameter estimation process itself (when gradient-based optimization methods are used), or may be developed explicitly (whether or not the parameter estimation is based on optimization).

3.1.1 Discussion Questions and Summary

Following the presentations, a number of questions were posed to the workshop participants to stimulate discussion.

Question 1.

Consider the following relationships:

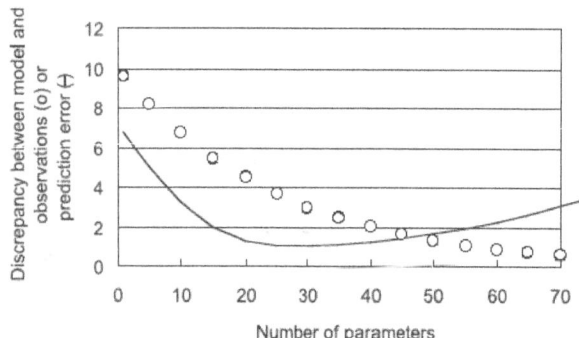

Source: M.C. Hill and C.R. Tiedeman, "Weighting observations in the context of calibrating groundwater models," *Calibration and Reliability in Groundwater Modelling: A Few Steps Closer to Reality* (Proceedings of ModelCARE'2002 (Prague, Czech Republic, 17–20 June 2002). IAHS Publ. no. 277, 2002.

A better model fit does not always lead to better predictions, particularly for out-of-sample predictions. What strategies can be used to drive the parameter estimation process toward the point of minimum prediction error?

Discussion. In response to this question, it was noted that the figure represents a multiobjective approach to calibration and that the figure implies multiple models (i.e., each with a different number of parameters). A suggestion was made to use model selection criteria to balance model fit and model complexity. Other criteria are also available to avoid fitting observation error. Several participants suggested that uncertainty estimates of the values plotted could be used to reach a decision about the appropriate level of model complexity. It was noted that in a real application the prediction error is unknown and that it is generally assumed the model structure is correct (or that one of a small set of considered model structures is correct); as a result, calibration should be viewed as more of a learning process. A related comment noted that a model doesn't need to predict reality to be useful. Another suggestion was the use of independent calibration and validation datasets, although it was pointed out that splitting a dataset may be unsuccessful when the model will be used to analyze the system under different conditions than those represented by the data. One participant noted that model challenges have been successful when conditions can be found under which the model fails; thus, the robustness of model results across a variety of conditions is important.

Question 2.

In "A comparison of seven geostatistically based inverse approaches to estimate transmissivities for modeling advective transport by groundwater flow" (Zimmerman et al., *Water Resources Research*, 34(6):1373–1413, 1998), the authors concluded the following:

> "It is disturbing to see that the available methods still do not adequately assess the uncertainty of the prediction." (pg. 1404)

> "The total uncertainty could therefore be better described by the results of the ensemble of several methods, as any one single method in general tends to underestimate the uncertainty." (pg. 1405)

Are these conclusions valid, in general? (How) can parameter estimation techniques be improved to better reflect actual uncertainty?

Discussion. A participant noted that the biggest problem for the inverse methods used in this paper was when the data were contrived to make the site non-Gaussian. In these circumstances, combining the appropriate geological elements with the models was essential to obtaining an accurate model. It was noted that the actual uncertainty can never be estimated; one can build confidence in a model, but cannot validate a model. A participant related an experience in which a site was modeled by a number of groups. The groups initially underestimated the uncertainty, but as they shared information between the groups, the uncertainty estimates increased. Thus, it is important to broaden both the number of models and the number of experts. Another participant related an experience in expert elicitation of uncertainties in which the degree of uncertainty was judged to be much larger by the outside academic experts than by the onsite experts. A related comment stressed the importance of peer review.

Question 3.

Should observation error be incorporated into the parameter estimation method? This includes emphasizing accurate observations and de-emphasizing inaccurate observations. Most people will say 'yes' to this, but in practice people often increase the weights of observations that provide a lot of information on estimated parameters, so that the weights indicate a greater degree of confidence in the measurement than is justifiable. Is this practice likely to produce more accurate predictions?

Discussion. There was some disagreement expressed about the subjective weighting of data in optimization-based parameter estimation. A participant stated that observation error and model error are reasons the model would not fit the data; it is thus useful to subjectively change the weights since they represent more than the observation error. Another participant commented that it is difficult to determine how to include model error in the weights and asked whether it would not be better to let the weight represent the observation error and thereby try to get some estimate of the model error. It was noted that observation errors themselves may be subjective. A participant asked what information was being added by subjectively adjusting the weights. Another pointed out that subjectively adjusting the weights introduces bias. A participant related an experience with a model in which one data point was felt to be less credible and was given less weight in the parameter estimation. The parameterized model was subsequently shown to violate a common interpretation of one of the boundary conditions, at least in part as a result of the data weighting, and was rejected by the regulatory body on that basis. Two years were then spent on data collection, which corroborated that the data initially felt to be less credible was true and should not have been deemphasized.

Question 4.

The Interagency Steering Committee on Radiation Standards is currently developing a database of parameter values and distributions for multimedia environmental modeling. What is the role in parameter estimation of such a database?

Discussion. A participant stated that an appropriate role for parameter databases is to provide prior distributions in a Bayesian sense. Another participant countered that the danger is that such priors may also end up serving as final distributions. It was suggested that this danger could be ameliorated by maximizing the uncertainty in the database values. It was questioned what parameter distributions in such a database represent: uncertainty or variability. If variability, is it variability of the mean over many sites, variability from individual to individual in a population, or some other quantity? One participant noted that although only physically based, complex models were discussed in this session, multimedia environmental models, as typically applied, are simpler, seldom calibrated, and may not use site-specific data. A participant responded that if multimedia environmental models are applied on a national scale, to tens or hundreds of sites, that it may not be practical to calibrate each one, to make an informed decision. In addition, reliance on regional or national databases may be required when site-specific measurements cannot be made. A comment noted the importance of good metadata to prevent data misuse. Another stated that every model has a mixture of data and that the nature of the data needs to be described. In some cases, parameters are less important to the predictions of interest and therefore don't require site-specific values; sensitivity analysis can help identify these parameters. A participant noted that generic parameter values can be useful to advance the calculations until further along in the analysis. It was suggested that an analysis similar to that conducted under the Prediction in Ungauged Basins program would be useful for other media (such as groundwater). Finally, it was stated that managers are interested in more than a single value for a parameter; they want measures of parameter uncertainty such as specific percentile values or bounding values.

3.1.2 Application Issues

Model-independent software is currently available for optimization-based parameter estimation, as represented by UCODE and PEST. These codes (or similar methods) have been applied to a wide variety of problems, including atmospheric, surface water, vadose zone, and ground-water models. Barriers to the application of optimization-based parameter estimation arise from two conditions.

One, there may be insufficient system response data to conduct the optimization. If the collection of this data is not feasible, then, as discussed above, *a priori* parameter estimates must be made by extrapolating from knowledge of parameters at other sites, by indirect estimation using relationships between parameter values and system characteristics that have been measured (or can be more easily measured than system response), and by direct measurement. Additional development in techniques, databases, and relationships between parameters and system characteristics are needed to improve the scientific basis for parameter values (and parameter probability distributions) estimated without a direct comparison between model predictions and observed system response.

Two, there may be sufficient data for optimization-based parameter estimation, but the additional analytical and computational effort required discourages or precludes its application. This is most likely to be an issue for models using complex representations of processes and detailed spatial or temporal resolution. Additional developments in parameter estimation methods for these complex models along with the software to implement them are needed.

3.1.3 Lessons Learned

The primary research emphasis in parameter estimation has been on development and improvement of optimization-based methods. These methods have been widely applied and codes are available for application with any model. Techniques to improve performance of the optimization-based methods are an emphasis of current research.

Applications in a variety of environments have demonstrated that calibration improves the performance of models. Nonetheless, there are a substantial number of applications of multimedia environmental models that do not use formal calibration methods to estimate parameters. There are a number of reasons why this may be the case. As mentioned above, system response data may not be available, or the application may be so computationally demanding that calibration is not feasible. In some cases, it may be that the model itself is not amenable to calibration because it does not predict measurable system response quantities. For example, the model may predict values averaged over a large domain, or it may only report derived values such as dose or risk. Comparatively little research has been directed at improving methods for *a priori* parameter estimation, including assessing the uncertainty of such estimates.

3.1.4 Research Needs

Research needs identified by the participants included the following:

- Appropriate regularization methods for highly parameterized models that best encapsulate the modeler's knowledge while providing numerical stability

- Predictive uncertainty analysis for highly parameterized models, including the combination of regularized inversion with Monte Carlo methods and efficient ways to approximate predictive confidence intervals without Monte Carlo analysis

- Methods to couple multi-criteria parameter estimation with probabilistic uncertainty analysis

- Algorithms for generation of alternative models

- Automation of model evaluation/comparison methods

- Additional field applications of novel sensitivity measures to moderately or highly nonlinear models and to highly parameterized models

- Consideration of conceptual model uncertainty as well as parameter uncertainty in sensitivity measures

- Improved methods for *a priori* parameter estimation through application of a wide variety of models to a wide range of data sets

- A detailed analysis of model process conceptualization and associated *a priori* parameter estimation methods (a focus of MOPEX)

- Improved models for adsorption (e.g., surface complexation), including development of parameter databases for these models

- Methods/applications to establish the predictive capabilities of improved adsorption models at the lab and field scales

- Methods to combine generic pedotransfer functions and site-specific information for the estimation of soil hydraulic properties

- Methods to include the effects of soil structure, soil chemistry, and clay mineralogy in pedotransfer functions

- Incorporation of additional parameter estimation and uncertainty analysis methods in the HYDRUS code

- Additional development of the JUPITER code including application codes

3.1.5   Conclusions

Developments in optimization-based parameter estimation have been sufficient that a number of codes are available that can be applied to the wide variety of models used by different Federal agencies. In addition, APIs under development, such as the JUPITER API discussed in this session and the COSU API discussed in Session 5, will improve the integration and comparison of models and existing parameter estimation tools and facilitate collaboration in the development of new tools. These developments should expand the set of modeling applications that use optimization-based parameter estimation methods and help to mature the science and technology.

Corresponding efforts to develop tools to facilitate *a priori* parameter estimation and to integrate such tools across model applications and federal agencies are limited. Existing databases of generic parameter values are often model-specific. In addition, there have been an insufficient number of applications in which it has been possible to evaluate the suitability of generic parameter databases and *a priori* parameter estimation methods. In this regard, it may be valuable to replicate the MOPEX experience in other media/applications.

Unsaturated Zone Parameter Estimation Using the HYDRUS and Rosetta Software Packages

Martinus Th. van Genuchten, Jirka Simunek, Marcel G. Schaap and Todd H. Skaggs

George E. Brown, Jr. Salinity Laboratory, USDA-ARS, Riverside, California,
rvang@ussl.ars.usda.gov, jiri.simunek@ucr.edu, mschaap@ussl.ars.usda.gov,
tskaggs@ussl.ars.usda.gov

The Salinity Laboratory has long developed and used parameter estimation codes to estimate a variety of soil hydraulic and solute transport parameters from laboratory and/or field experimental data. Much of our earlier work focused on estimating parameters in analytical solute transport models (Skaggs et al., 2002), such as the physical (mobile-immobile) and chemical (two-site) nonequilibrium models embedded in the CFITIM (van Genuchten, 1981) and CXTFIT (Parker and van Genuchten, 1984; Toride et al., 1995) codes. Recently a Windows-based version (STANMOD, Simunek et al., 1999b) of these and related one- and multidimensional analytical transport models became available.

Using parameter optimization techniques for estimating the unsaturated soil hydraulic properties became popular in the mid 1980s (e.g., Kool et al., 1985), initially in conjunction with mostly one- and multi-step outflow experiments. Such optimizations require numerical solutions of the governing Richards equation for variably saturated flow because of the highly nonlinear relationships between the water content, the hydraulic conductivity, and the pressure head (or suction). As more flexible and comprehensive numerical programs such as the HYDRUS codes (Simunek et al., 1998, 1999a; Rassam et al., 2003) became available, these studies were extended to analyses of upward flux or head-controlled infiltration experiments (including tension infiltrometry), evaporation methods, or any other experiment involving some appropriate combination of water flow and solute transport data. In this paper, we briefly review the main features of the HYDRUS codes and their utility for estimating soil hydraulic and solute transport parameters. Also, as an alternative to using HYDRUS for site-specific parameter estimation studies, we briefly summarize the Rosetta code for estimating the unsaturated soil hydraulic parameters and their uncertainty in a more generic manner from soil texture and related surrogate data that are often available. Details of these and other models discussed in this paper can be found at the Web Site of the Salinity Laboratory (www.ussl.ars.usda.gov/models/models.htm).

The Windows-based modular HYDRUS-1D and HYDRUS-2D software packages may be used to address one- and two-dimensional flow and contaminant transport problems, respectively. The HYDRUS codes use the Richards equation for variably-saturated flow and Fickian-based advection-dispersion equations for both heat and solute transport. The flow equation considers water uptake by plant roots, as well as hysteresis in the unsaturated soil hydraulic properties. The solute transport equations include provisions for nonlinear sorption, one-site and two-site non-equilibrium transport, dual-porosity media involving mobile and immobile water, and the transport of solute decay chains. The software packages come with Levenberg-Marquardt type nonlinear parameter optimization modules to allow estimation of a variety of soil hydraulic and solute transport parameters from experimental data. Unknown hydraulic parameters may be estimated from observed water contents, pressure heads, and/or boundary fluxes during transient flow by numerical inversion of the Richards equation. Additional retention or hydraulic conductivity data, as well as a penalty function for constraining the optimized parameters to remain in some feasible region (Bayesian estimation) can be optionally considered. The procedure similarly permits solute transport and/or reaction parameters to be estimated from observed concentrations and related data.

Agricultural applications of HYDRUS include irrigation and drainage design, salinization of irrigated lands, pesticide leaching and volatilization, virus transport in the subsurface, and analysis of riparian systems. Typical non-agricultural problems include the design of radioactive waste disposal sites, contaminant leaching from landfills, design and analysis of capillary barriers, transport and degradation of chlorinated hydrocarbons, and recharge from deep vadose zones. Any of these applications, in principle, may involve parameter estimation. Several strategies can be followed for this purpose. First, one could use water flow information only (e.g., pressure heads and/or fluxes) to estimate the soil hydraulic parameters, followed by estimation of the transport parameters using information from the transport part of the experiment (e.g., solute concentrations). Alternatively, combined water flow and transport information can be used to estimate soil hydraulic and solute transport parameters in a sequential manner. Finally, combined water flow and transport information can be used to simultaneously estimate both the soil hydraulic and solute transport parameters. This last approach is the most beneficial since it uses crossover effects between state variables and parameters, and takes advantage of all available information since concentrations are a function of water flow. Several studies have shown that simultaneous estimation of hydraulic and transport properties yields smaller estimation errors for model parameters than sequential estimation.

Even with the use of parameter estimation software, appropriate experiments for determining the unsaturated soil hydraulic properties can be very time-consuming and costly. One alternative is to use pedotransfer functions (PTFs) to indirectly estimate the hydraulic properties from more easily measured and/or readily available data such as soil texture and bulk density. We developed a Windows-based software package, Rosetta, for this purpose. The PTFs in Rosetta are based on a combined bootstrap-neural network procedure to predict water retention parameters and the saturated and unsaturated hydraulic conductivity, as well as their probability distributions. The PTFs were calibrated on a large number of soil hydraulic data sets derived from three different databases, including the UNSODA unsaturated soil hydraulic database developed at the Salinity Laboratory (Nemes et al., 2001). Rosetta offers a hierarchical set of five PTFs to predict van Genuchten-Mualem type hydraulic parameters depending upon available information, from limited data (soil textural class only) to more extensive data (texture, bulk density, and one or two water retention points). One attractive feature of Rosetta is that it provides uncertainties in its parameter estimates (Figure 1). Uncertainty estimates are generated with the bootstrap method and are given as standard deviations around the estimated hydraulic parameters (Schaap et al., 2001). The uncertainties, which depend upon the invoked PTF model and its input data, are useful in cases where few or no hydraulic data are available. They are particularly useful for risk-based simulations of water flow and solute transport.

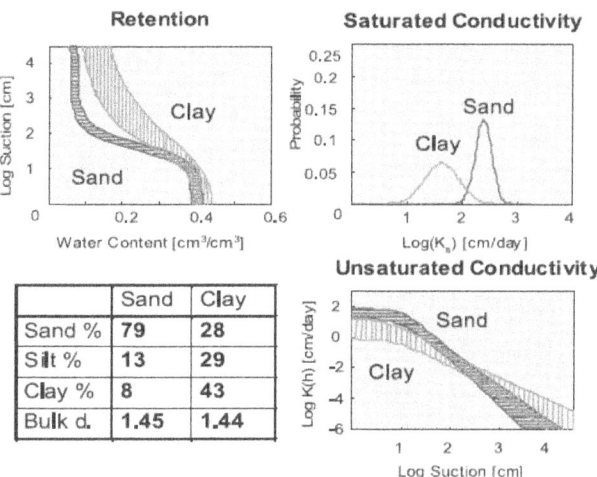

Figure 1. Examples of 90% confidence intervals generated with Rosetta for water retention and the unsaturated hydraulic conductivity for two soils.

References

Nemes, A., M.G. Schaap, F.J. Leij, and J.H.W. Wösten. 2001. Description of the unsaturated soil hydraulic database UNSODA, version 2.0. J. Hydrol. 251:151-162.

Parker, J.C., and M. Th. van Genuchten. 1984. Determining Transport Parameters from Laboratory and Field Tracer Experiments. Bull. 84-3, Virginia Agric. Exp. Sta., Blacksburg, VA, 91 p.

Rassam, D., J. Simunek, and M. Th. van Genuchten. 2003. Modelling Variably-Saturated Flow with HYDRUS-2D. ND Consult, Brisbane, Australia, 275 p.

Schaap, M.G., F.J. Leij, and M. Th. van Genuchten. 2001. Rosetta: A computer program for estimating soil hydraulic parameters with hierarchical pedotransfer functions. J. Hydrol. 251:163-176.

Simunek, J., K. Huang, M. Šejna, and M. Th. van Genuchten. 1998. The HYDRUS-1D Software Package for Simulating the One-Dimensional Movement of Water, Heat and Multiple Solutes in Variably-Saturated Media, Version 1.0. IGWMC-TPS-70, Int. Ground Water Modeling Center, Colorado School of Mines, Golden, CO., 186 p.

Simunek, J., K. Huang, M. Šejna, and M. Th. van Genuchten. 1999a. The HYDRUS-2D Software Package for Simulating Two-Dimensional Movement of Water, Heat and Multiple Solutes in Variably-Saturated Media. Version 2.0. IGWMC-TPS-53, Int. Ground Water Modeling Center, Colorado School of Mines, Golden, CO, 251 p.

Simunek, J., M. Th. van Genuchten, M. Šejna, N. Toride, and F.J. Leij. 1999b. The STANMOD software package for evaluating solute transport in porous media using analytical solutions of the convection-dispersion equation. Version 1.0. IGWMC-TPS-71, Int. Ground Water Modeling Center, Colorado School of Mines, Golden, CO. 32 p.

Skaggs, T.H., D.B. Jaynes, R.G. Kachanoski, P.J. Shouse, and A.L. Ward. 2002. Solute transport: Data analysis and parameter estimation. In: J.J. Dane and G.C. Topp (eds.), Methods of Soil Analysis, Part 4. Physical Methods, pp. 1403-1434, Soil Sci. Soc. Am., Inc., Madison, WI.

Toride, N., F.J. Leij, and M. Th. van Genuchten. 1995. The CXTFIT code for estimating transport parameters from laboratory or field tracer experiments. Version 2.0. Research Report No. 137, U.S. Salinity Laboratory, USDA, ARS, Riverside, CA, 121 p.

van Genuchten, M. Th. 1981. Nonequilibrium transport parameters from miscible displacement experiments. Research Report No. 119, U.S. Salinity Laboratory, Riverside, CA. 88 p.

Parameter Estimation and Predictive Uncertainty Analysis for Ground and Surface Water Models Using PEST

John Doherty

Watermark Numerical Computing,

University of Queensland, Brisbane, Australia

PEST is a software package designed to undertake model-independent parameter estimation and predictive uncertainty analysis. The cornerstone of its model independence is its ability to communicate with a model through the model's own input and output files. Thus, a model does not need to be cast as a subroutine to be used with PEST. Nor does the model need to be a single executable program. In fact, the "model" can be a batch or script file comprised of the model itself (or a number of models) together with appropriate pre- and post-processors; this allows enormous flexibility in the design of the parameter estimation and predictive analysis process. To take full advantage of this, PEST is accompanied by a suite of utility software designed to optimize its use in the ground and surface water modeling contexts. Not only does this software carry out important pre-and post-processing tasks; specific members of this utility suite are able to automate construction of an entire PEST input dataset based on calibration designs involving considerable complexity.

PEST's parameter estimation algorithm is based on the Gauss-Marquardt-Levenberg (GML) method. However, considerable effort has been devoted to making the version of this algorithm implemented in PEST as robust as possible. To further enhance PEST's performance in difficult calibration settings, PEST includes functionality for manual and "automatic" user intervention; this allows selective removal of troublesome parameters (normally insensitive parameters) from the parameter estimation process. PEST also allows the imposition of bounds on parameter values. Bounds enforcement is undertaken by selective, temporary, "freezing" of parameters; the order in which parameters are frozen depends on their trajectories with respect to the GML-calculated parameter upgrade vector, and to the objective function gradient vector.

Versions of PEST from 5.0 onwards have included sophisticated regularization functionality. The use of regularized inversion allows the estimation of many more parameters than would otherwise be possible in a numerically stable manner. Furthermore, if regularization conditions are properly imposed, estimated parameter values "make sense" in the context simulated by the model. In groundwater modeling, regularized inversion allows the use of complex spatial parameterization schemes. For example, parameters can be based on pilot points or even on individual model cells. Regularization constraints can also be flexible, being based on smoothness, minimum curvature, "heterogeneity focusing" or any of a variety of other methodologies. See Doherty (2002) for further details; see also Figure 1, which shows the estimated hydraulic conductivity distribution over the domain of the Eastern Snake Plain groundwater model. (This model is being built by personnel from the University of Idaho, Idaho Falls.) An interesting variant of the use of regularized inversion is its combination with stochastic field generation to undertake "calibration-constrained Monte-Carlo analysis" in which regularization constraints enforce minimized deviation from a stochastic "seed field." Current PEST development work includes the introduction of more flexible storage and data handling capabilities for regularized inversion based on very large parameter sets; the use of prediction-constrained regularized inversion to assign probability ranges to different characterizations of subsurface hydraulic heterogeneity; and the development of optimal regularization schemes for use in different geological contexts. Part of this work has been made possible by Tom Clemo from Boise State University who has recently developed an efficient adjoint state solver for MODFLOW.

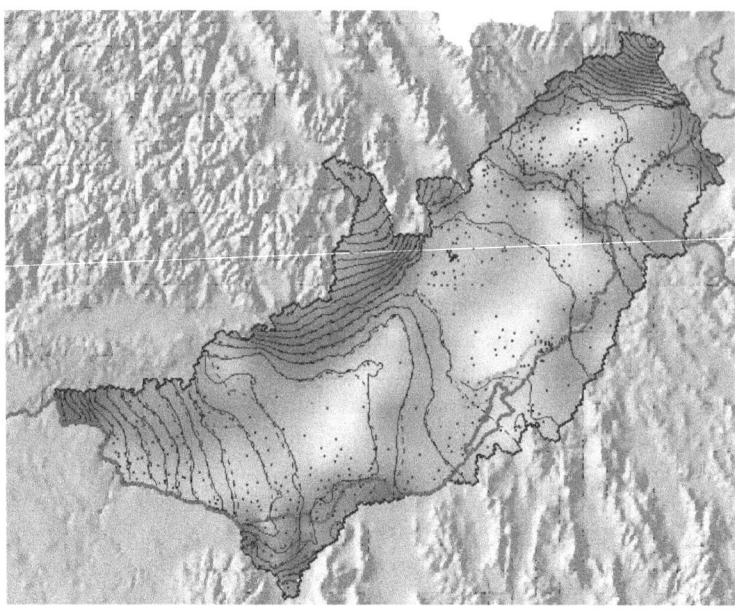

Figure 1. Estimated hydraulic conductivity distribution for Eastern Snake Plain, Idaho.

Use of PEST in the surface water modeling context has also relied on regularized inversion as a mechanism for accommodating the highly parameterized nature of such models. In a typical application of a surface water quality model such as HSPF, parameter uniqueness is rare. The situation is compounded where submodels for multiple land-uses and soil types must be calibrated on the basis of flow and quality data acquired at a location downstream from all of these simulated systems. Here the challenge facing surface water modelers is to assimilate (sometimes vague) information regarding relative parameter values in different sub-watersheds (based on implicit relationships between these parameters and the real-world system which they represent), while at the same time respecting the fact that the "lumped" nature of parameters used by these models makes adherence to such relationship tenuous. PEST's regularization functionality is invoked to provide a good fit between model outputs and field data while adhering to preferred parameter values (and/or relationships between parameter values) to the maximum extent possible without compromising this fit.

Success in calibrating surface water quantity and quality models necessitates the construction of a multi-component objective function. Different aspects of a flow or constituent time series are rich in information pertaining to different model parameters. One of the challenges that must be faced in optimizing the design of the inverse problem in this context is the "distillation" of these different aspects, incorporating each of them into the objective function with sufficient weight to be "seen" by PEST. To date, this "distillation" process has involved the inclusion of entities such as flow volumes, flow statistics, sediment rating curves, and even digitally filtered flows, in the overall objective function. The result has been much greater numerical stability on the part of PEST, and greater uniqueness in parameter estimates (with greater confidence in these estimates as a result). See Doherty and Johnston (2003) for further details.

Despite advances such as these in estimating parameters for surface water models, parameter and predictive nonuniqueness is nevertheless a major problem in this type of modeling. As is being increasingly noted (see for example NRC, 2001), it is incumbent on modelers to analyze predictive uncertainty as a routine part of the model calibration and deployment process. PEST is able to accommodate this imperative through its predictive analysis functionality. The algorithmic basis of this capability is presented in Vecchia and Cooley (1987). It should be noted that PEST's ability to maximize or minimize a key model prediction while maintaining calibration constraints is not based on any linearity assumption. The user simply provides PEST with the objective function at which the model is deemed to be "uncalibrated" (at a certain probability level); PEST will then maintain this constraint (and thus maintain the model in a "calibrated state") while maximizing or minimizing the identified prediction. Parameter reality can be maintained in this process through imposition of parameter bounds (see above) – either directly on each parameter, or on the relationships between parameters.

References

Doherty, J., 2002. Groundwater model calibration using Pilot Points and Regularization. *Ground Water*. Vol 41 (2): 170–177

Doherty, John, and John M. Johnston, 2003. Methodologies for calibration and predictive analysis of a watershed model, *J. American Water Resources Association*, 39(2):251–265.

National Research Council. 2001. Assessing the TMDL Approach to Water Quality Management. National Academy Press. Washington, DC, 109pp.

Vecchia, A.V. and Cooley, R.L., 1987. Simultaneous confidence and prediction intervals for nonlinear regression models with application to a groundwater flow model. Water Resour. Res. 23 (7):1237–1250.

PEST and its utility software can be downloaded from the following Web site: http://www.sspa.com/pest

A Priori Parameter Estimation: Issues and Uncertainties

George Leavesley

U.S. Geological Survey, Denver, Colorado

A major difficulty in the use of distributed-parameter models is the general lack of objective methods to estimate the distributed values of parameters. Calibration techniques are typically used to compensate for various sources of uncertainty in these estimates. However, the transferability of calibrated parameters is often an issue due to the incorporation of a variety of error sources in the fitted values and the general over-parameterization of many distributed-parameter models. In addition, the application of these models to complex problems, such as ungauged basins, or assessing the impact of land-use and climate change, is further limited because there are typically no measures of system response available against which to calibrate. Estimating parameters where optimization is not possible, and addressing the over-parameterization problem by minimizing the number of parameters to be fitted, necessitate the use of parameter-estimation methods that rely on the use of measurable climatic and basin characteristics.

The development of methodologies to relate various model process parameters to basin characteristics has been conducted by a number of disciplines in the field of hydrology. Studies at the point and plot scale have typically been used to define these relations. However, the application and evaluation of such techniques at larger scales have been limited. The increasing availability of high-resolution spatial and temporal data sets of climatic and basin characteristics now provides the opportunity to investigate parameter-estimation techniques at large scales and over a wide range of climatic and physiographic regions.

Cooperative research efforts have been initiated among a variety of national and international organizations to take advantage of these data sources and to begin addressing the issues of *a priori* parameter estimation and the uncertainty associated with the use of *a priori* parameter estimates. These research programs include the Model Parameter Estimation Experiment (MOPEX) project (http://www.nws.noaa.gov/oh/mopex) and the Predictions in Ungauged Basins (PUB) project (http://iahs.info). A discussion of the science issues associated with these types of research efforts and preliminary results of the MOPEX program are presented.

Multi-Objective Approaches for Parameter Estimation and Uncertainty

Luis A. Bastidas

Civil and Environmental Engineering, Utah Water Research Laboratory
Utah State University, Logan, Utah

The goal of parameter estimation is to achieve a reduction in model uncertainty by efficiently extracting information contained in observational data. Several complementary criteria should be used to extract information about different model components or parameters, thereby enhancing the overall identifiability of the model. The traditional multi-criteria approach has been to select several different criteria and then merge them together into a single function for optimization. However, there is a significant advantage to maintaining the independence of the various performance criteria and that a full multi-criteria optimization should be performed to identify the entire set of Pareto optimal solutions. In particular, the multi-criteria approach offers a way of emulating the Manual-Expert calibration of employing a number of complementary ways of evaluating the model performance, compensating for various kinds of model and data errors, and extracting greater amounts of information from the data. This presentation will explore the major issues regarding the approach and propose specific questions for further research.

Using Sensitivity Analysis in Model Calibration Efforts

Claire R. Tiedeman, U.S. Geological Survey, Menlo Park, California

Mary C. Hill, U.S. Geological Survey, Boulder, Colorado

In models of natural and engineered systems, sensitivity analysis can be used to assess relations among system state observations, model parameters, and model predictions. The model itself links these three entities, and model sensitivities can be used to quantify the links. Sensitivities are defined as the derivatives of simulated quantities (such as simulated equivalents of observations, or model predictions) with respect to model parameters. We present four measures calculated from model sensitivities that quantify the observation-parameter-prediction links and that are especially useful during the calibration and prediction phases of modeling. These four measures are composite scaled sensitivities (CSS), prediction scaled sensitivities (PSS), the value of improved information (VOII) statistic, and the observation prediction (OPR) statistic. These measures can be used to help guide initial calibration of models, collection of field data beneficial to model predictions, and recalibration of models updated with new field information. Once model sensitivities have been calculated, each of the four measures requires minimal computational effort.

We apply the four measures to a three-layer MODFLOW-2000 (Harbaugh et al., 2000; Hill et al., 2000) model of the Death Valley regional ground-water flow system (DVRFS), located in southern Nevada and California. D'Agnese et al. (1997, 1999) developed and calibrated the model using nonlinear regression methods. Figure 1 shows some of the observations, parameters, and predictions for the DVRFS model. Observed quantities include hydraulic heads and spring flows. The 23 defined model parameters include hydraulic conductivities, vertical anisotropies, recharge rates, evapotranspiration rates, and pumpage. Predictions of interest for this regional-scale model are advective transport paths from potential contamination sites underlying the Nevada Test Site and Yucca Mountain.

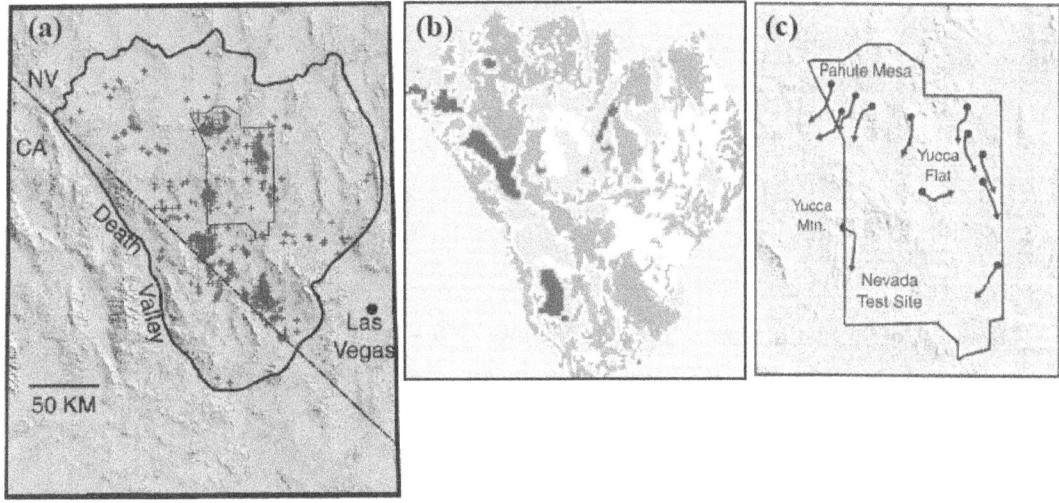

Figure 1: (a) Hydraulic-head observation locations, (b) distribution of hydraulic conductivity parameters in model layer 1, and (c) advective transport predictions, for the three-layer DVRFS model.

Composite scaled sensitivities (CSS) address the observation-parameter link. CSS identify the support provided by observations towards estimating the value of each model parameter, and can be used to define sets of parameters to estimate during calibration (Hill, 1998). CSS are commonly calculated throughout the calibration process, starting with uncalibrated models. For the DVRFS, CSS calculated for an initial model with few parameters helped guide introduction of additional parameters. CSS calculated for the final model helped identify nine parameters that could be estimated by regression, given the available hydraulic-head and flow observations (D'Agnese et al., 1997, 1999).

Prediction scaled sensitivities (PSS) and the value of improved information (VOII) statistic address the parameter-prediction link. PSS and the VOII statistic are generally calculated using a calibrated model, and identify parameters that are important to the model predictions (Tiedeman et al., 2003). PSS are a fairly simple measure of parameter importance, and are calculated as a scaled version of the sensitivity of a predicted value with respect to a model parameter. The VOII statistic is a more complex measure that accounts for parameter correlations. It quantifies the decrease in prediction uncertainty that would be produced by obtaining improved field information on one or more model parameters. The PSS and VOII results can help guide field collection of new hydrogeologic data for improving the model predictions and reducing prediction uncertainty. This can be achieved by collecting field data about parameters identified as most important to the predictions, incorporating these data into an updated model, and recalibrating the model.

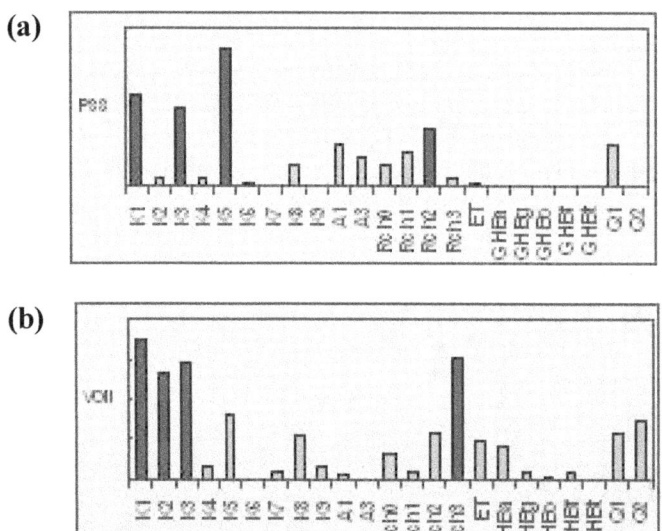

Figure 2: Evaluation of the importance of the DVRFS model parameters to a predicted advective transport path on Yucca Flat, using (a) prediction scaled sensitivities (PSS) and (b) the value of improved information (VOII) statistic.

Figure 2 shows results of using the PSS and VOII statistics to evaluate the importance of DVRFS model parameters to a predicted advective transport path on Yucca Flat. These results indicate that some of the important parameters represent flow system attributes that are distant from the path (Tiedeman et al., 2003). This transport path remains entirely in hydraulic conductivity zone K1 and is overlain by recharge zone Rch0, yet additional hydraulic conductivity and recharge parameters clearly rank as important to this prediction.

The Observation Prediction (OPR) statistic addresses the observation-prediction link. It is generally calculated using a calibrated model, and measures the change in model prediction uncertainty that would be produced if an observation were added to or removed from an existing monitoring network (Hill et al., 2001). The OPR statistic can be used to guide removal of less important observations from an existing monitoring network, by identifying observations that, if omitted, would not substantially increase prediction uncertainty. It can also be used to guide future data collection, by identifying locations where collection of additional observations would produce the greatest reductions in prediction uncertainty.

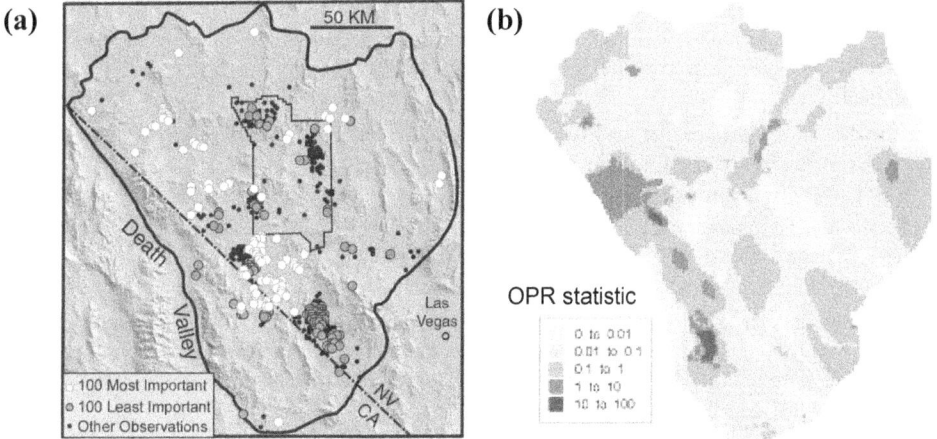

Figure 3: Evaluation of the importance of (a) existing DVRFS hydraulic-head observations and (b) potential new head observation locations in model layer 1 to predicted advective-transport paths, using the observation prediction (OPR) statistic.

Figure 3 shows results of applying the OPR statistic to evaluate hydraulic-head observations for the DVRFS model. Analysis of the existing hydraulic-head monitoring network showed that many unimportant observations are in areas of high observation density, and thus could be removed from the network without diminishing its broad geographic coverage (Hill et al., 2001). Evaluation of potential new observations showed that the most important new observation locations are mainly in areas of high hydraulic gradient.

The CSS, PSS, VOII, and OPR results were obtained for the DVRFS model using sensitivities produced by MODFLOW-2000. However, the measures can be determined for any application model for which sensitivities can be calculated, such as by using UCODE (Poeter and Hill, 1998) or PEST (Doherty, 2003).

The validity of the CSS, PSS, VOII, and OPR results assumes that the model is accurate and is linear with respect to the model parameters. Evaluation of the DVRFS model indicates that it is reasonably accurate, but that it is nonlinear. However, the degree of nonlinearity is mild enough for the four measures calculated from model sensitivities to be useful.

References

D'Agnese, F.A., C.C. Faunt, A.K. Turner, and M.C. Hill, 1997, Hydrogeologic evaluation and numerical simulation of the Death Valley regional ground-water system, Nevada and California, using geoscientific information systems: U.S. Geological Survey Water-Resources Investigations Report 96-4300, 124 p.

D'Agnese, F.A., C.C. Faunt, M.C. Hill, and A.K. Turner, 1999, Death Valley regional ground-water flow model calibration using optimal parameter estimation methods and geoscientific information systems: Advances in Water Resources, v. 22, no. 8, p. 777–790.

Doherty, John, 2003, PEST v. 7.0, Watermark Computing, http://www.sspa.com/PEST/index.html.

Harbaugh, A. W., E.R. Banta, M.C. Hill, and M.G. McDonald, 2000, MODFLOW-2000, The U.S. Geological Survey modular ground-water model, Users guide to modularization concepts and the ground-water flow process: U.S. Geological Survey Open-File Report 00-92, 121 p, http://water.usgs.gov/nrp/gwsoftware/modflow2000/modflow2000.html.

Hill, M.C., 1998, Methods and guidelines for effective model calibration: U.S. Geological Survey Water-Resources Investigations Report 98-4005, 90 p, http://pubs.water.usgs.gov/wri984005/.

Hill, M.C., E.R. Banta, A.W. Harbaugh, and E.R. Anderman, 2000, MODFLOW 2000, The U.S. Geological Survey modular ground-water model, User's guide to the observation, sensitivity, and parameter-estimation processes: U.S. Geological Survey Open-File Report 00-184, 209 p, http://water.usgs.gov/nrp/gwsoftware/modflow2000/modflow2000 html.

Hill, M.C., D.M. Ely, C.R. Tiedeman, F.A. D'Agnese, C.C. Faunt, and G.A. O'Brien, 2001, Preliminary evaluation of the importance of existing hydraulic-head observation locations to advective-transport predictions, Death Valley regional flow system, California and Nevada: U.S. Geological Survey Water-Resources Investigations Report 00-4282, 82p, http://water.usgs.gov/pubs/wri/wri004282/.

Poeter, E.P. and M.C. Hill, 1998, Documentation of UCODE, A computer code for universal inverse modeling: U.S. Geological Survey Water-Resources Investigations Report 98-4080. 116p, http://water.usgs.gov/software/ucode.html.

Tiedeman, C.R., M.C. Hill, F.A. D'Agnese, and C.C. Faunt, 2003, Methods for using groundwater model predictions to guide hydrogeologic data collection, with application to the Death Valley regional ground-water flow system, Water Resour. Res., 39(1), 1010, DOI:10.1029/2001WR001255.

JUPITER Project—Merging Inverse Problem Formulation Technologies

Mary Hill[1], Eileen Poeter[2], John Doherty[3], Edward R. Banta[4], and Justin Babendreier[5]

[1]U.S. Geological Survey, Boulder, Colorado, USA, mchill@usgs.gov
[2]Colorado School of Mines, Golden, Colorado, USA, epoeter@mines.edu
[3]Watermark Computing and University of Queensland, Brisbane, Australia
[4]U.S. Geological Survey, Lakewood, Colorado, USA, erbanta@usgs.gov
[5]Environmental Protection Agency, Georgia, USA, Babendreier.Justin@epamail.epa.gov

The JUPITER (Joint Universal Parameter IdenTification and Evaluation of Reliability) project seeks to enhance and build on the technology and momentum behind two of the most popular sensitivity analysis, data assessment, calibration, and uncertainty analysis programs used in environmental applications: PEST (Doherty, 1994, 2002) and UCODE (Poeter and Hill, 1998). These programs are universal in that they can be applied to any computer model; both have very flexible methods for interacting with application models through ASCII files. PEST and UCODE have enjoyed substantial success. Their future, however, depends on their transition into a well-designed, flexible Application Programming Interface (API) that will support new ways of interacting with application models and new, sophisticated capabilities. Much of the technology incorporated in UCODE and PEST has been investigated thoroughly enough that its strengths, weaknesses, and advantageous uses are fairly well known. The frontier of model calibration and associated analysis methods includes pursuits that will benefit from a stable, modularly programmed, full-featured, well-designed, thoroughly documented foundation. JUPITER will provide that foundation for the PEST and UCODE developers, the work of our contemporaries and, we hope, the work of coming generations.

There are two ongoing phases of the JUPITER project. The first phase is the development of the JUPITER API, which will include (1) conventions for program input and output and internal data production and consumption, and (2) subroutines that support commonly used calculations and manipulations. The JUPITER API takes advantage of the Framework for Risk Analysis in Multimedia Environmental Systems (FRAMES) API (Castleton, 2003; Babendreier, 2003) and the Uncertainty Analysis/Sensitivity Analysis/Parameter Estimation (UA/SA/PE) API (ISCMEM API Workgroup, 2003). The second phase is development of the first applications of the JUPITER API, J_UCODE and J_PEST. J_UCODE will replace the existing UCODE, and will have enhanced capabilities for generating and investigating alternative conceptual models. This enhancement of UCODE will be the focus of part of this talk. J_PEST will replace PEST, including the nonlinear confidence intervals that form the basis of its prediction analyzer, and a capability for using regularization methods that allow parameter values to be defined at virtually every basic entity of a numerical model (generally, this would be a finite-difference cell or a finite element).

The JUPITER API will provide the opportunity for users to better evaluate data sets using JUPITER application codes (application codes that use the JUPITER API) to readily (1) experiment with a number of techniques for generating conceptual models (e.g., geostatistical methods, geologic process modeling, upscaling); (2) compare alternative parameter estimation algorithms (e.g., J_PEST and J_UCODE); (3) "mine" results from various conceptual models for model evaluation, ranking and multi-model inferential analysis, as well as use these results to evolve the conceptual model (e.g., unreasonable parameter-value estimates provide clues to hydrogeologic structure; residual bias provides clues to data bias); and (4) assess data needs to improve the calibration in light of the predictions. These tools will be useful in the conceptual model development and evaluation procedure suggested by Neuman and Wierenga (2003).

Future work includes developing utility codes; creating JUPITER application codes; improving model generation algorithms; automating model evaluation; and encouraging community contributions.

References

Babendreier, J.B., 2003, The Multimedia, Multipathway, Multireceptor Risk Assessment Modeling System (FRAMES-3MRA Version 1.0) Documentation. Volume IV: Evaluating Uncertainty and Sensitivity. Draft SAB Review Report: EPA530/D/03/001d. Office of Solid Waste and Office of Research and Development, Washington DC.. http://www.epa.gov/ceampubl/mmedia/3mra/index.htm. See also Volumes I, II, III, and V: EPA530/D/03/001a:b:c:e.

Castleton, K.J., 2003, Framework for Risk Analysis in Multimedia Environmental Systems (FRAMES) Version 2.0 Application Programming Interface (API). http://mesa6.mesastate.edu/~kcastlet/API/download/

Doherty, J., 1994, PEST: Model Independent Parameter Estimation. Watermark Numerical Computing. Australia. http://www.sspa.com/pest/

Doherty, J., 2002, PEST-ASP: Version 5 of PEST. Watermark Numerical Computing. Australia. http://www.sspa.com/pest/

ISCMEM API Workgroup, 2003, UA/SA/PE API. The reader is referred to the last day's Workshop Session Theme: Towards Development of a Common Software Application Programming Interface (API) for Uncertainty, Sensitivity, and Parameter Estimation Methods & Tools.

Neuman, S.P. and P.J. Wierenga, 2003, A Comprehensive Strategy of Hydrogeologic Modeling and Uncertainty Analysis for Nuclear Facilities and Sites, NUREG/CR-6805, U.S. Nuclear Regulatory Commission, Washington, DC.

Poeter, E.P., and M.C. Hill, 1998, Documentation of UCODE: A Computer Code for Universal Inverse Modeling, Water Resources Investigations Report 98-4080, U.S. Geological Survey, Lakewood, Colorado. http://typhoon.mines.edu/software/igwmcsoft/

Simulated Contaminant Plume Migration:
The Effects of Geochemical Parameter Uncertainty

L.J. Criscenti, R.T. Cygan, M. Siegel, M. Eliassi

Sandia National Laboratories, Albuquerque, New Mexico 87185-0750

There is little consensus on how chemical reactions and reaction parameters should be determined for field applications. In particular, several models for contaminant adsorption onto mineral surfaces are used to describe both laboratory and field observations. Contaminant adsorption is dependent on numerous variables that are difficult to quantify including the surface area and site density of the adsorbing minerals, the characteristics of the boundary layer between the mineral surface and bulk solution, and both the structure and composition of the adsorbing species. Models to describe contaminant adsorption range from the strictly empirical distribution coefficient (Kd) model to sophisticated multisite surface complexation models that provide a mechanistic model for the adsorption of a specific ion to specific mineral surface sites. The simpler Kd models are valid only under the conditions of measurement. The surface complexation models are valid over a larger range of environmental conditions, but have, in general, only been parameterized for very simple laboratory systems that may not be representative of the field. In order to use these models to describe field observations, assumptions must be made regarding the dominant adsorption reactions.

In one study, broadly based on the hydrogeology and mineralogy of the Naturita uranium mill tailings site, we assume all uranium is removed from the tailings leachate through adsorption onto smectite, an abundant clay mineral present in the field. Experimental results show that uranium adsorbs to specific surface sites on both the basal planes and edges of smectite. We chose to model this adsorption using a two-site surface complexation model. Because uranium adsorbs predominantly to the aluminum edge surface sites [>(e)AlOH], we elected to examine uncertainty only in the equilibrium constants associated with these sites. We used one- and two-dimensional reactive-transport models to numerically examine variations in predicted contaminant migration due to uncertainty in the adsorption constants. Using the Latin Hypercube Sampling method, one hundred pairs of adsorption constant (log K) values are selected for the surface species >(e)AlO- and >(e)AlOUO2+, from normal distributions of each log K. One-dimensional simulations were performed to examine the removal of adsorbed uranium from contaminated soil by the influx of rainwater. The simulation results can be identified by two distinct groups of uranium breakthrough curves. In the first group, the breakthrough curves exhibit a classical sigmoidal shape whereas in the second group the breakthrough curves display higher uranium concentrations in solution over greater distances and times. These two groups are clearly separated by two different ranges of log K >(e)AlO- values or two different ranges for the smectite point of zero charge. Two-dimensional simulations were performed to examine the migration of uranium from a tailings site into an uncontaminated aquifer. Sensitivity analysis shows that, for this set of simulations, the shape and size of the predicted contaminant plumes are functions of log K >(e)AlOUO2+. The uncertainties associated with the geochemical parameters yielded larger variations in calculated contaminant migration than uncertainties in longitudinal dispersivity or aquifer heterogeneity as defined by a random porosity distribution.

Funding provided by the U.S. Nuclear Regulatory Commission, Office of Nuclear Regulatory Research. Sandia is a multiprogram laboratory operated by Sandia Corporation, a Lockheed-Martin company, for the U.S. Department of Energy under contract DE-AC04-94AL85000.

Reference: Criscenti, L.J., M. Eliassi, R.T. Cygan, C.F. Jové Colón (2002). Effects of Adsorption Constant Uncertainty on Contaminant Plume Migration: One- and Two-Dimensional Numerical Studies. NUREG/CR-6780.

Impacts of Sensitive Parameter Uncertainties on Dose Impact Analyses for Decommissioning Sites

Boby Abu-Eid and Mark Thaggard

Division of Waste Management
Office of Nuclear Material Safety and Safeguards
U.S. Nuclear Regulatory Commission
Washington, DC 20555-0001

Dose impact analyses are conducted by NRC's Staff and licensees for decommissioning of facilities contaminated with residual radioactivity to demonstrate compliance with the dose criteria in 10 CFR Part 20, Subpart E [e.g., 0.25 mSv (25 mrem) per year for unrestricted release or 1/5 mSv (100/500 mrem) per year for restricted release]. The dose analysis results are commonly used to establish radionuclide derived concentration guideline level (DCGL), for site release, corresponding to the dose limit. NRC's Staff developed guidance documents and codes/models to enable a probabilistic approach for calculation of the Total Effective Dose Equivalent (TEDE) to the average member of the critical group (e.g., NUREGs-1727, -1757 Vol. 2, and codes/models documented in NUREG/CR-6676, -6692, -6697, and -5512).

The parameters used in the dose analysis typically pertain to (a) release of the residual radioactive source material to environmental media (e.g., to air, surface/subsurface soil, surface water and groundwater) and to the biota; (b) transport of radionuclide through environmental pathways that lead to dispersion of radionuclides in environmental media; (c) human exposure through pathways such as direct exposure, air/dust inhalation, ingestion of drinking water, and biotic contamination; and (d) dose conversion factors for calculation of the TEDE to the exposed average member of the critical group or cancer risk factors. The generic dose modeling input parameters may be grouped into three categories: (a) physical parameters (P) that are dependent on the source, its location, and the geological or physical characteristics of the site, and also independent of the group receptors (e.g., distribution coefficients, hydraulic conductivity); (b) behavioral parameters (B) that are dependent on the receptor behavior and the scenario employed in the dose analysis (e.g., occupancy parameters, diet consumption); and (c) metabolic parameters (M) that are dependent on the receptor and independent of the scenario (e.g., inhalation rates, milk consumption).

The NRC developed common tools, codes, and models to help staff and licensees conduct probabilistic dose analysis. For example, Lookup Tables and DandD Version 2.1 code were developed for screening analysis (NUREG-1727, NUREG-1757, and NUREG/CR-5512 Vol. 1, 2, and 3). For screening analysis, the most sensitive and problematic parameters include the resuspension factor, the mass loading factor for foliar deposition, and bio-transfer factors. NRC Staff is developing new approaches to minimize excessive conservatism in the distributions and selection of these parameters (e.g., draft NUREG-1720). For generic site-specific analysis, the NRC developed probabilistic RESRAD >6.0 and RESRAD-BUILD >3.0 codes and established template distributions for most sensitive parameters. The staff also developing stylized calculation approaches for complex decommissioning sites using models such as GEN-II, RESRAD-OFFSITE, MEPAS, and platforms such as FRAMES and GOLDSIM. The stylized calculation is also used for the generic environmental impact assessment. Further, the NRC will participate in development of probabilistic RESRAD-OFFSITE for potential use for complex sites with offsite releases and use in environmental impact assessments.

The staff adopted a simple approach for initial identification of sensitive parameters. The approach is essentially based on the relevancy of a parameter to the dose calculation and the degree of parameter influence on the peak dose calculations. A quantity called the normalized dose difference (NDD) is used as an indicator for initial selection of sensitive parameters:

$$NDD = (D_{high} - D_{low})/D_{base} \times 100\%$$

Where $(D_{high} - D_{low})$ is the range of the peak dose calculated when the parameter is set at its high and low values, and D_{base} is the peak dose when the parameter is set at its base value. The base value uses a well studied default parameter value and a mixture of radionuclides sources with a concentration of 1 pCi/g for each, in a contaminated zone area of 2,400 m² and a contamination depth of 0.15 m. The radionuclide mixture includes radionuclides: Co-60, Sr-90, Cs-137, Ra-226, Th-230, U-238, Pu-239, and Am-241. The peak dose was calculated for the different parameter ranges and correlated with the base peak value. Table 1 shows examples of the most sensitive input physical parameters for RESRAD code and the degree of sensitivity using the NDD indicator.

The current probabilistic dose analysis methodology, using common probabilistic codes/models, involves: (a) sampling of sensitive parameters from parameter distribution inputs using simple random, "Monte Carlo" based sampling,on the requested number of observations by the user or Latin Hypercube Sampling (LHS) where one sample is obtained from each non-overlapping area of equal probability; (b) use of the parameter statistical distributions; NRC Staff currently uses 40 default radionuclide independent parameters' statistical distributions (e.g., the erosion rate, inhalation rate, and thickness of the unsaturated zone) and five radionulcide dependent parameters (e.g., distribution coefficients, transfer factors). For parameters that do not have default distributions, or for modifying a distribution, staff may choose from more than 30 statistical distributions (e.g., continuous: uniform, loguniform, triangular, normal, exponential, beta, and gamma; and discrete: Poisson, Geometric, Binomial, Negative Binomial, and Hypergeometric). The most common distributions comprise Lognormal (19), Triangular (19), Uniform/Loguniform (14), Normal 9, and Empirical (5). Site-specific distributions could be established based on available relevant data and performance of Bayesian statistical analysis (e.g., through assessment of likelihood, obtaining posterior distribution, and estimation of posterior means). The values of posterior means can be entered into the code for statistical parameter values. (c) Use of "Input Rank Correlation"; staff usually provides inputs of a relationship between two parameters using a correlation coefficient with a range of -1 for a strong negative correlation (e.g., porosity and bulk density) and +1 for strong positive correlation (e.g., porosity and effective porosity). The output correlations used to examine the sensitivity of input parameters include: (i) Partial Correlation Coefficient (PCC), which indicates how linear the correlation is; (ii) Standard Regression Coefficient (SRC) which indicates how sensitive a parameter in a linear model; (iii) Partial Rank Correlation Coefficient (PRCC), which is typically used for nonlinear models and multiple parameters; and (iv) Standard Rank Regression Coefficient (SRRC) which is used to indicate sensitivity of the parameter.

In summary, the probabilistic approach for uncertain sensitive parameters requires:

(1) Examination of the parameters influencing the dominant pathways (e.g., pathways with significant contribution to the dose output) and examining the causes of influence. For example high K_d values of U-238 in the CZ may highly increase the dose related to plant ingestion and low K_d values may highly increase the peaking time dose related to drinking water ingestion. Staff also employs "scatter plots" to identify the probabilistic variables that have the most influence on the dominant pathway dose and on the overall dose. For example plots of K_d values of U in the UZ and the SZ versus the dose from ingestion of water and the dose from all pathways show a dose variation between 0.01 and 35 mrem/yr. The scatter plots of the plant transfer factor for the Sr show a dose range of 0.01 to 140 mrem/yr.

(2) Study of the distributions of parameters and the interrelationships between the influential factors to assess the range and interrelationship between the probabilistic variables that have the most influence on each other and on the dose. Evaluation of the relationship between similar parameters is quite common. For example a direct high rank correlation was noted for K_d s of the contaminated zone (CZ), the unsaturated zone (UZ), and the saturated zone (SZ).

(3) Evaluation and potential development of a probability distribution appropriate for the particular site, if necessary. This may decrease the variability of the dose output

(4) Performing "Linear Regression" between the output dose and the input parameters. It is recommended to use output raw data if linearly related, or ranked data if the output is monotonically related to the inputs; or use of coefficients of determination if the relationship is not known (e.g., K_d value vs. dose from plant ingestion or doses from plant ingestion versus U plant transfer factor).

(5) Increasing number of observations and number of repetitions will generally reduce uncertainties in the output dose distribution; however, calculation time will increase.

The general outputs in the dose analysis include (a) Peak of the Mean (POM), dose for each repetition and the time POM dose occurs; (b) the Mean of the Peak (MOP) dose; (c) the percentile dose and the Cummulative Distribution Function (CDF) of the peak dose; (d) scatter plots of the dose vs. input parameter; (e) the mean dose of summed all pathways. The end point for the deterministic dose analysis is the peak dose or soil guideline derived using the peak dose. For the probabilistic analysis, the endpoint is the distribution of the peak doses selected at different percentiles or peak of the mean dose at various times along the time horizon

In conclusion, sensitive parameters may impact the dose result by a factor that may reach 1–2 orders of magnitude or more. It is recommended to assess sensitive parameters based on site-specific conditions and examine the causes of their impacts on the dominant pathway doses and the overall output dose value. Parameter uncertainties could be reduced significantly through establishing interrelationships between the influential factors, or parameters, and through assessment of the ranges between the probabilistic variables that have the most influence on each other and the dose.

Table 1: Examples of Most Sensitive Physical Parameters Using NDD* Indicator

Parameter	Radionuclide NDD							
	Co-60	SR-90	Cs-137	Ra-226	Th-230	U-238	Pu-239	Am-241
External γ Shielding	54	0	48	7	7	0	0	7
Cover Depth and Density of Cover Material	98	6	92	11	159	1	9	51
	250	0	85	2	0	0	0	0.1
Density of CZ	26	1.4	23	56	74	62	58	0.2
Distribution Coefficients (CZ, UZ, SZ)	0.9	3	6	0.1	51	94	95	0.1
SZ Hydraulic Conductivity and effective porosity	0	0	0	0	0	114	117	0
	0	0	0	0	0	146	150	0
UZ Thickness	0	0	0	0	0	96	96	0
Depth of Roots	3	253	18	10	15	0	0	131
Transfer Factors for Plants, Meat, and Milk	1	89	13	42	56	0	0	480
	5	101	42	2	5	3	1	36
	3	180	55	8	10	30	0	5
Mass Loading for Inhalation	0	0	0	0	2	0	0	35

* NDD = $(D_{high} - D_{low})/D_{base}$ x 100% where the $(D_{high} - D_{low})$ is the range of the peak dose calculated when the parameter is set at its high and low values, and D_{base} is the peak dose when the parameter is set at its base value.

4

SESSION 2:

SENSITIVITY ANALYSIS APPROACHES, APPLICATIONS, AND LESSONS LEARNED — *IDENTIFICATION OF RESEARCH NEEDS*

Overview and Summary

Editors: Sitakanta Mohanty and Thomas Nicholson

The session had six invited presenters. Most of these presentations focused on uncertainty and sensitivity analyses related to parameters. Two presentations highlighted the use of sensitivity/ uncertainty analysis in risk analysis and the decision-making process. Two presentations dealt with groups of parameters or components of the system being modeled. The general methods discussed at the workshop included Fractional Factorial Design (FFD) (Andres), Sampling-Based Methods (Helton), a combined Regional Sensitivity Analysis (RSA) and a Tree-Structured Density Estimation method (TSDE) (Osidele and Beck), and Global Sensitivity approaches (Saltelli). An application of several different methods to a large and complex model was presented to illustrate the areas of applicability and general deficiencies (Mohanty). Another presentation highlighted the broader issues related to the implementation of uncertainty/sensitivity analysis in risk assessment (Frey). The following is a summary of sensitivity analysis applications, lessons learned, and identification of research needs discussed during these presentations.

4.1.1 Discussion Questions

The following questions were posed by the session moderator and rapporteur to facilitate discussion:

1. What are the unique issues to pay attention to, or key challenges to overcome, while carrying out sensitivity analysis in surface-water or ground-water flow and transport problems?

2. Which sensitivity analysis methods are most promising for surface-water and ground-water flow and transport applications?

3. What are the essential informational and data needs for implementing and demonstrating these methods to surface-water and/or ground-water flow and transport analyses?

4. How does sensitivity analysis relate to parameter estimation and uncertainty analysis?

(Although these questions were posed to the audience for discussion, much of the following narrative, as did the session, focused primarily on question 1 and partially on question 4. Although the audience did not directly comment on questions 2 and 3, these questions were acknowledged as where the Working Group would like to proceed in order to create the "tool box" as identified by George Leavesley, WG Chair, in his opening remarks.)

4.1.2 Discussion Summary

Uncertainty and sensitivity analyses are needed to identify areas for improvement and to provide input to prioritize resource allocation and develop action plans. They are also needed to reduce unnecessary regulatory burden. They drive development of a common understanding in a multi-disciplinary environment.

Uncertainty analysis is carried out with the intention of revealing where major sources of uncertainties are and how they affect risk estimates. In uncertainty analysis, the question for which we seek answer is what is the uncertainty in analysis results given the uncertainty in analysis inputs?

Sensitivity analysis identifies factors (i.e., events, processes, components, designs, and model limitations) contributing most to system behavior. In sensitivity analysis, the question for which we seek answer is how important are the individual elements of the input vector with

67

respect to response of the analysis results. As an example, differential-based sensitivity analysis identifies where a small input perturbation has a large effect on system response. To contrast with uncertainty analysis, sensitivity analysis is the mapping of inferences onto assumptions, while uncertainty analysis is the converse process. It was noted that, in the uncertainty analysis framework, the importance of input element uncertainty to the analysis results variability is studied, which is referred to as uncertainty importance analysis.

4.1.3 Application Issues

Uncertainty Analysis Considerations: In quantitative risk assessment, an important and potentially expensive part of uncertainty analysis is the characterization of uncertainty in the input parameters [i.e., represented via probability density functions (pdfs)]. Care is needed in constructing the probability density functions because in the analysis that forms the basis for important decisions, the probability density functions typically influence both uncertainty and sensitivity analysis results. But in practice, the rigor (the level of care and effort) with which the probability density functions are identified depends on the purpose of analysis and time and resources available. From a model computation standpoint, the most demanding part of the analysis is the propagation of sampled data through the analysis.

Sensitivity Analysis Considerations: Sensitivity analysis of model parameter sets can be expensive, especially for a model that has a large parameter set (i.e., hundreds or thousands). For example, the cost of sensitivity analysis using the FFD approach can grow as $O(N^2)$, where N is the number of parameters and O is the order of magnitude, and $O(N)$ is the time for setting up the run and executing it. Thus, from a computational standpoint, the goal should be to minimize $O(N^2)$ as much as possible.

A variety of sensitivity analysis techniques should be used to gain insights into the system model. In addition to sensitivity analysis with respect to individual parameters, it should also be carried out with respect to the complement of models and sub-models, groups of parameters, and subsystems (e.g., components and processes) to gain better understanding of system's behavior. Different parameter transformations of a single output variable can also yield different groups of influential parameters (i.e., significant relative impact on model outputs). Therefore, such parameter transformations can be used to further understand the model behavior.

4.1.4 Lessons Learned

Observations made by the various experts include the following:

- Whatever the method one uses, it is important that the framing of the analysis should be defensible for the modeler and meaningful to its users.

- The target of interest in sensitivity analysis should not be the model output per se, but to answer the central question for which the model was formulated. Similarly, the relevancy of the model is not the focus, but the relevancy of the model conclusions addressing the problem being solved.

- Sensitivity analysis should be used prior to model development, during model development, and when the model is used during analysis.

- Sensitivity, uncertainty importance (i.e., sensitivity analyses in the presence of uncertainty), and robustness analyses are key components of probabilistic risk assessment.

- Systematic model simplification (i.e., model abstraction) which still retains the key processes, uncertainties, and variability is important to practical probabilistic risk assessment.

- Parametric sensitivity analysis provides useful risk insights, but alternative approaches are also needed to understand "which" parameters showed up as important and "why" they showed up as important. Explicit statements on model assumptions, limitations, data, accuracy, subjectivities, and processes are needed to derive risk significance from the uncertainty and sensitivity analyses.

- Sensitivity analysis can be a valuable tool in building confidence in the model and the computer codes that embed these models. Therefore, software and model confidence building should be kept in mind while planning and performing risk assessments.

- In spite of current advances, the state-of-the-science has not matured to the point of quantitatively deriving risk significance from uncertainty/sensitivity analyses as input to final decision making.

Recommendations on analysis methods focused on sensitivity analysis. The use of global sensitivity methods (as opposed to the "one-factor-at-a-time" analysis methods) was emphasized, although most methods currently in use are some sort of global sensitivity methods, thought not explicitly recognized. Global sensitivity analysis is defined as the study of how the uncertainty in the output of a model (numerical or otherwise) can be appointed to different sources of uncertainty in the model input. Saltelli advocated the use of variance-based sensitivity measures. These measures are concise and easy to understand and communicate. The application of these variance-based measures reduces the problem to an elementary test for linear models. This approach relates to the popular method of Morris. Saltelli also advocated the use of sensitivity methods in the Monte-Carlo filtering family.

4.1.5 Research Needs

For additional investigation on uncertainty analysis, identified research needs fell into two broad categories: (a) alternative to conventional uncertainty representations (e.g., evidence theory and possibility theory), and (b) education. Educating the importance of uncertainty/sensitivity analysis was highlighted, specifically in the areas of (1) separating and identifying epistemic and aleatory uncertainties, (2) designing and implementing risk/performance analysis involving large and complex systems, and (3) substantiating conservative assumptions.

Several recommendations were made concerning future research needs in *sensitivity analysis*.

- Explore new sensitivity analysis procedures such as developing methods for non-parametric regression, 2-D Kolmogorov-Smirnov test, tests for non-monotone relations, tests for nonrandom patterns, and complete variance decomposition.

- In the Design-of-Experiment (DOE) approach, study how in practice, the number and influence of influential parameters vary with the number of realizations.

- Develop a standard interface for generating experimental designs and using them to drive model sensitivity analysis.

- Explore the use of factors prioritization, factors fixing, factors mapping, and variance cutting approaches.

In addition to identifying modeling-specific research needs, some subjective research needs were also identified. In *uncertainty analysis*, there is a need for more complete reporting of information regarding variability and uncertainty in data (e.g., systematically report mean, standard deviation, sample size). There is a need for credible (i.e., accepted) procedures for documenting expert judgment on data uncertainty and variability, suitable to a particular assessment objective. For *sensitivity analysis*, the analysis challenges are not always related simply to research, rather to determine whether the analyses will be requested by and later accepted by stakeholders and decision-makers. Methodological research studies need to be carried out that will identify problems of medium- to long-term policy interest, or recurring problems. These studies should focus on methodological gaps that prevent appropriate assessments necessary to a good decision-making process.

There was a concern that uncertainty and sensitivity analysis methods could be incorrectly used to make a case in favor or against a project. Therefore, there is a need to develop guidance documents (with experts' involvement or endorsement) that will provide the practitioners with the knowledge of what is available, and the context where the methods can be used (i.e., when to use them, and how to use them). Documentation of case studies where there have been successful communication of uncertainty and sensitivity analyses to support actual decisions between analysts and decision-makers should be made.

4.1.6 Conclusions

Recent developments illustrate the tremendous need for implementing quantitative uncertainty and sensitivity analyses. Numerous methods exist in the literature for conducting such analyses. These methods are available and effective for use today as evidenced by the technical literature, software, affordable computational resources, tested practices, and ease of communication. However, the greatest challenge remaining is the process of utilizing these analyses in decision-making.

A gap does remain in public education of the utility and implementation of uncertainty/sensitivity analysis methods in the decision-making process. In solving most problems, or in the decision-making process, subjective (qualitative) engineering judgment will continue to temper quantitative results in determining risk significance.

Global Sensitivity Analysis: Novel Settings and Methods

A. Saltelli

European Commission,
Joint Research Centre of Ispra, Italy
andrea.saltelli@jrc.it

This presentation wants to be an introduction to global sensitivity analysis (SA). Its ambition is to target an audience unfamiliar with global sensitivity analysis, and to give practical hints about the associated advantages and the effort needed.

We shall review some techniques for sensitivity analysis, including those that are not global, by applying them to a simple example. This will give the audience a chance to contrast each method's result against its own expectation of what the sensitivity pattern for the simple model should be. We shall also try to relate the discourse on the relative importance of model input factors to specific questions, such as *"Which of the uncertain input factor(s) is so non-influential that we can safely fix it/them?"* or *"If we could eliminate the uncertainty in one of the input factors, which factor should we choose to reduce the most the variance of the output?"*

In this way, the selection of the method for sensitivity analysis will be put in relation to the framing of the analysis and to the interpretation and presentation of the results. The choice of the output of interest will be discussed in relation to the purpose of the model-based analysis.

The example will show how the methods are applied in a way that is unambiguous and defensible, so as to making the sensitivity analysis an added value to model-based studies or assessments. This shall be put into context in relation with the post-modern critique of the use of mathematical models.

When discussing sensitivity with respect to factors, we shall interpret the term "factor" in a very broad sense: a factor is anything that can be changed prior to the execution of the model, possibly from a prior or posterior, continuous or discrete distribution. A factor can either be stochastically or epistemically uncertain. Factors can be "triggers," used to select one versus another model structure, one mesh size versus another, or altogether different conceptualisations of the system. The links with established Bayesian model averaging procedures will be mentioned.

The main methods that we present in this lecture are all related with one another, and are the method of Morris for factors' screening and the variance-based measures. All are model-free, in the sense that their application does not rely on special assumptions on the behavior of the model (such as linearity, monotonicity and additivity of the relationship between input factor and model output). Monte Carlo filtering will also be mentioned in relation to a framing of the analysis where the question of interest is *"Which of the input factors is mostly responsible for producing realizations of the output of interest in a given target region?"*

Finally, a set of worked examples (e.g., application of global sensitivity analysis to real models) is mentioned briefly to illustrate possible useful practices, and reference is given to the existing literature on the subject. Some most common pitfalls will be mentioned as well.

The presentation takes inspiration from a primer on sensitivity analysis that will appear for Wiley and Sons Publishers in early 2004.

Sampling-Based Methods for Uncertainty and Sensitivity Analysis

Jon C. Helton

Sandia National Laboratories
Albuquerque, New Mexico 87185-0779

Sampling-based approaches to uncertainty and sensitivity analysis are both effective and widely used [1-4]. Analyses of this type involve the generation and exploration of a mapping from uncertain analysis inputs to uncertain analysis results. The underlying idea is that analysis results $y(x) = [y_1(x), y_2(x), ..., y_{nY}(x)]$ are functions of uncertain analysis inputs $x = [x_1, x_2, ..., x_{nX}]$. In turn, uncertainty in x results in a corresponding uncertainty in $y(x)$. This leads to two questions: (i) What is the uncertainty in $y(x)$ given the uncertainty in x?, and (ii) How important are the individual elements of x with respect to the uncertainty in $y(x)$? The goal of uncertainty analysis is to answer the first question, and the goal of sensitivity analysis is to answer the second question. In practice, the implementation of an uncertainty analysis and the implementation of a sensitivity analysis are very closely connected on both a conceptual and a computational level.

Implementation of a sampling-based uncertainty and sensitivity analysis involves five components: (i) Definition of distributions $D_1, D_2, ..., D_{nX}$ that characterize the uncertainty in the components $x_1, x_2, ..., x_{nX}$ of x, (ii) Generation of a sample $x_1, x_2, ..., x_{nS}$ from the x's in consistency with the distributions $D_1, D_2, ..., D_{nX}$, (iii) Propagation of the sample through the analysis to produce a mapping $[x_k, y(x_k)]$, $k = 1, 2, ..., nS$, from analysis inputs to analysis results, (iv) Presentation of uncertainty analysis results (i.e., approximations to the distributions of the elements of y constructed from the corresponding elements of $y(x_k)$, $k = 1, 2, ..., nS$), and (v) Determination of sensitivity analysis results (i.e., exploration of the mapping $[x_k, y(x_k)]$, $k = 1, 2, ..., nS$). The five preceding steps will be discussed and illustrated with results from a performance assessment for the Waste Isolation Pilot Plant (WIPP) [5-7].

Definition of the distributions $D_1, D_2, ..., D_{nX}$ that characterize the uncertainty in the components $x_1, x_2, ..., x_{nX}$ of x is the most important part of a sampling-based uncertainty and sensitivity analysis as these distributions determine both the uncertainty in y and the sensitivity of y to the elements of x. The distributions $D_1, D_2, ..., D_{nX}$ are typically defined through an expert review process [8-11], and their development can constitute a major analysis cost. A possible analysis strategy is to perform an initial exploratory analysis with rather crude definitions for $D_1, D_2, ..., D_{nX}$ and use sensitivity analysis to identify the most important analysis inputs; then, resources can be concentrated on characterizing the uncertainty in these inputs and a second presentation or decision-aiding analysis can be carried with these improved uncertainty characterizations.

Several sampling strategies are available, including random sampling, importance sampling, and Latin hypercube sampling [12, 13]. Latin hypercube sampling is very popular for use with computationally demanding models because its efficient stratification properties allow for the extraction of a large amount of uncertainty and sensitivity information with a relatively small sample size. In addition, effective correlation control procedures are available for use with Latin hypercube sampling [14, 15]. The popularity of Latin hypercube sampling recently led to the original article being designated a *Technometrics* classic in experimental design [16].

Propagation of the sample through the analysis to produce the mapping $[\mathbf{x}_k, \mathbf{y}(\mathbf{x}_k)]$, $k = 1, 2, \ldots$, nS, from analysis inputs to analysis results is often the most computationally demanding part of a sampling-based uncertainty and sensitivity analysis. The details of this propagation are analysis specific and can range from very simple for analyses that involve a single model to very complicated for large analyses that involve complex systems of linked models [7, 17].

Presentation of uncertainty analysis results is generally straightforward and involves little more than displaying the results associated with the already calculated mapping $[\mathbf{x}_k, \mathbf{y}(\mathbf{x}_k)]$, $k = 1, 2, \ldots$, nS. Presentation possibilities include means and standard deviations, density functions, cumulative distribution functions (CDFs), complementary cumulative distribution functions (CCDFs), and box plots [2, 13]. Presentation formats such as CDFs, CCDFs, and box plots are usually preferable to means and standard deviations because of the large amount of uncertainty information that is lost in the calculation of means and standard deviations.

Determination of sensitivity analysis results is usually more demanding than the presentation of uncertainty analysis results due to the need to actually explore the mapping $[\mathbf{x}_k, \mathbf{y}(\mathbf{x}_k)]$, $k = 1, 2, \ldots$, nS, to assess the effects of individual components of $\mathbf{x}$ on the components of $\mathbf{y}$. Available sensitivity analysis procedures include examination of scatterplots, regression analysis, correlation and partial correlation analysis, stepwise regression analysis, rank transformations to linearize monotonic relationships, identification of nonmonotonic patterns, and identification of nonrandom patterns [2-4, 18, 19].

Sampling-based uncertainty and sensitivity analysis is widely used, and as a result, is a fairly mature area of study. However, there still remain a number of important challenges and areas for additional study. For example, there is a need for sensitivity analysis procedures that are more effective at revealing nonlinear relations than those currently in use. Possibilities include procedures based on nonparametric regression [20-22], the two-dimensional Kolmogorov-Smirnov test [23-25], tests for nonmonotone relations [26], tests for nonrandom patterns [27-31], and complete variance decomposition [32, 33]. As another example, sampling-based procedures for uncertainty and sensitivity analysis usually use probability as the model, or representation, for uncertainty. However, when limited information is available with which to characterize uncertainty, probabilistic characterizations can give the appearance of more knowledge than is really present. Alternative representations for uncertainty such as evidence theory and possibility theory merit consideration for their potential to represent uncertainty in situations where little information is available [34, 35]. Finally, a significant challenge is the education of potential users of uncertainty and sensitivity analysis about (i) the importance of such analyses and their role in both large and small analyses, (ii) the need for an appropriate separation of aleatory and epistemic uncertainty in the conceptual and computational implementation of analyses of complex systems [36-40], (iii) the need for a clear conceptual view of what an analysis is intended to represent and a computational design that is consistent with that view [41], and (iv) the importance of avoiding deliberately conservative assumptions if meaningful uncertainty and sensitivity analysis results are to be obtained.

References

1. Iman, R.L., 1992. "Uncertainty and Sensitivity Analysis for Computer Modeling Applications," *Reliability Technology - 1992, The Winter Annual Meeting of the American Society of Mechanical Engineers, Anaheim, California, November 8-13, 1992.* Eds. T.A. Cruse. Vol. 28. New York, NY: American Society of Mechanical Engineers, Aerospace Division. 153–168.

2. Helton, J.C., 1993. "Uncertainty and Sensitivity Analysis Techniques for Use in Performance Assessment for Radioactive Waste Disposal," *Reliability Engineering and System Safety.* Vol. 42, no. 2-3, pp. 327–367.

3. Hamby, D.M., 1994. "A Review of Techniques for Parameter Sensitivity Analysis of Environmental Models," *Environmental Monitoring and Assessment.* Vol. 32, no. 2, pp. 135–154.

4. Blower, S.M., and H. Dowlatabadi. 1994. "Sensitivity and Uncertainty Analysis of Complex Models of Disease Transmission: an HIV Model, as an Example," *International Statistical Review*. Vol. 62, no. 2, pp. 229-243.

5. Helton, J.C., D.R. Anderson, H.-N. Jow, M.G. Marietta, and G. Basabilvazo. 1999. "Performance Assessment in Support of the 1996 Compliance Certification Application for the Waste Isolation Pilot Plant," *Risk Analysis*. Vol. 19, no. 5, pp. 959–986.

6. Helton, J.C. 1999. "Uncertainty and Sensitivity Analysis in Performance Assessment for the Waste Isolation Pilot Plant," *Computer Physics Communications*. Vol. 117, no. 1-2, pp. 156–180.

7. Helton, J.C., and M.G. Marietta. 2000. "Special Issue: The 1996 Performance Assessment for the Waste Isolation Pilot Plant," *Reliability Engineering and System Safety*. Vol. 69, no. 1-3, pp. 1–451.

8. Hora, S.C., and R.L. Iman. 1989. "Expert Opinion in Risk Analysis: The NUREG-1150 Methodology," *Nuclear Science and Engineering*. Vol. 102, no. 4, pp. 323–331.

9. Thorne, M.C., and M.M.R. Williams. 1992. "A Review of Expert Judgement Techniques with Reference to Nuclear Safety," *Progress in Nuclear Safety*. Vol. 27, no. 2-3, pp. 83–254.

10. Budnitz, R.J., G. Apostolakis, D.M. Boore, L.S. Cluff, K.J. Coppersmith, C.A. Cornell, and P.A. Morris. 1998. "Use of Technical Expert Panels: Applications to Probabilistic Seismic Hazard Analysis," *Risk Analysis*. Vol. 18, no. 4, pp. 463–469.

11. McKay, M., and M. Meyer. 2000. "Critique of and Limitations on the use of Expert Judgements in Accident Consequence Uncertainty Analysis," *Radiation Protection Dosimetry*. Vol. 90, no. 3, pp. 325–330.

12. McKay, M.D., R.J. Beckman, and W.J. Conover. 1979. "A Comparison of Three Methods for Selecting Values of Input Variables in the Analysis of Output from a Computer Code," *Technometrics*. Vol. 21, no. 2, pp. 239–245.

13. Helton, J.C., and F.J. Davis. 2003. "Latin Hypercube Sampling and the Propagation of Uncertainty in Analyses of Complex Systems," *Reliability Engineering and System Safety*. Vol. 81, no. 1, pp. 23–69.

14. Iman, R.L., and W.J. Conover. 1982. "A Distribution-Free Approach to Inducing Rank Correlation Among Input Variables," *Communications in Statistics: Simulation and Computation*. Vol. B11, no. 3, pp. 311–334.

15. Iman, R.L., and J.M. Davenport. 1982. "Rank Correlation Plots for Use with Correlated Input Variables," *Communications in Statistics: Simulation and Computation*. Vol. B11, no. 3, pp. 335–360.

16. Morris, M.D., 2000. "Three Technometrics Experimental Design Classics," *Technometrics*. Vol. 42, no. 1, pp. 26–27.

17. Breeding, R.J., J.C. Helton, E.D. Gorham, and F.T. Harper. 1992. "Summary Description of the Methods Used in the Probabilistic Risk Assessments for NUREG-1150," *Nuclear Engineering and Design*. Vol. 135, no. 1, pp. 1–27.

18. Helton, J.C., and F.J. Davis. 2000. "Sampling-Based Methods for Uncertainty and Sensitivity Analysis," *Sensitivity Analysis*. A. Saltelli, K. Chan, and E.M. Scott (eds). New York, NY: Wiley. pp. 101–153.

19. Helton, J.C., and F.J. Davis. 2002. "Illustration of Sampling-Based Methods for Uncertainty and Sensitivity Analysis," *Risk Analysis*. Vol. 22, no. 3, pp. 622-691.

20. Hastie, T.J., and R.J. Tibshirano. 1990. *Generalized Additive Models*. London: Chapman & Hall.

21. Simonoff, J.S. 1996. *Smoothing Methods in Statistics*. New York: Springer-Verlag.

22. Bowman, A.W., and A. Azzalini. 1997. *Applied Smoothing Techniques for Data Analysis*. Oxford: Clarendon.

23. Peacock, J.A. 1983. "Two-Dimensional Goodness-Of-Fit Testing in Astronomy," *Monthly Notices of the Royal Astronomical Society*. Vol. 202, no. 2, pp. 615–627.

24. Fasano, G., and A. Franceschini. 1987. "A Multidimensional Version of the Kolmogorov-Smirnov Test," *Monthly Notices of the Royal Astronomical Society*. Vol. 225, no. 1, pp. 155–170.

25. Garvey, J.E., E.A. Marschall, and R. Wright, A. 1998. "From Star Charts to Stoneflies: Detecting Relationships in Continuous Bivariate Data," *Ecology*. Vol. 79, no. 2, pp. 442–447.

26. Hora, S.C., and J.C. Helton. 2003. "A Distribution-Free Test for the Relationship Between Model Input and Output when Using Latin Hypercube Sampling," *Reliability Engineering and System Safety*. Vol. 79, no. 3, pp. 333–339.

27. Ripley, B.D. 1979. "Tests of "Randomness" for Spatial Point Patterns," *Journal of the Royal Statistical Society*. Vol. 41, no. 3, pp. 368–374.

28. Diggle, P.J., and T.F. Cox. 1983. "Some Distance-Based Tests of Independence for Sparsely-Sampled Multivariate Spatial Point Patterns," *International Statistical Review*. Vol. 51, no. 1, pp. 11–23.

29. Zeng, G., and R.C. Dubes. 1985. "A Comparison of Tests for Randomness," *Pattern Recognition*. Vol. 18, no. 2, pp. 191–198.

30. Assunçao, R. 1994. "Testing Spatial Randomness by Means of Angles," *Biometrics*. Vol. 50, pp. 531-537.

31. Kleijnen, J.P.C., and J.C. Helton. 1999. "Statistical Analyses of Scatterplots to Identify Important Factors in Large-Scale Simulations, 1: Review and Comparison of Techniques," *Reliability Engineering and System Safety*. Vol. 65, no. 2, pp. 147–185.

32. Saltelli, A., S. Tarantola, and K.P.-S. Chan. 1999. "A Quantitative Model-Independent Method for Global Sensitivity Analysis of Model Output," *Technometrics*. Vol. 41, no. 1, pp. 39–56.

33. Li, G., C. Rosenthal, and H. Rabitz. 2001. "High-Dimensional Model Representations," *The Journal of Physical Chemistry*. Vol. 105, no. 33, pp. 7765–7777.

34. Klir, G.J., and M.J. Wierman. 1999. *Uncertainty-Based Information*, New York, NY: Physica-Verlag.

35. Helton, J.C., J.D. Johnson, and W.L. Oberkampf. 2004. "An Exploration of Alternative Approaches to the Representation of Uncertainty in Model Predictions," *Reliability Engineering and System Safety*. Vol. 85(1-3), pp. 39–71.

36. Apostolakis, G. 1990. "The Concept of Probability in Safety Assessments of Technological Systems," *Science*. Vol. 250, no. 4986, pp. 1359–1364.

37. Helton, J.C. 1994. "Treatment of Uncertainty in Performance Assessments for Complex Systems," *Risk Analysis*. Vol. 14, no. 4, pp. 483–511.

38. Hoffman, F.O., and J.S. Hammonds. 1994. "Propagation of Uncertainty in Risk Assessments: The Need to Distinguish Between Uncertainty Due to Lack of Knowledge and Uncertainty Due to Variability," *Risk Analysis*. Vol. 14, no. 5, pp. 707–712.

39. Paté-Cornell, M.E. 1996. "Uncertainties in Risk Analysis: Six Levels of Treatment," *Reliability Engineering and System Safety*. Vol. 54, no. 2-3, pp. 95–111.

40. Helton, J.C. 1997. "Uncertainty and Sensitivity Analysis in the Presence of Stochastic and Subjective Uncertainty," *Journal of Statistical Computation and Simulation*. Vol. 57, no. 1-4, pp. 3–76.

41. Helton, J.C. 2001. "Mathematical and Numerical Approaches in Performance Assessment for Radioactive Waste Disposal: Dealing with Uncertainty," *Etude pour la Faisabilité des Stochazes de Déchets Radioactifs, Actes des Journées Scientifiques ANDRA, Nancy, 7, 8, et 9 decembré 1999*. Les Ulis cedex A, France: EDP Sciences. 59–90.

Uncertainty and Sensitivity Analysis for Environmental and Risk Assessment Models

H. Christopher Frey

Department of Civil, Construction, and Environmental Engineering
North Carolina State University
Raleigh, North Carolina 27695-7908
frey@eos.ncsu.edu

This talk provides an overview of research in the areas of uncertainty and sensitivity analyses and regarding future research needs in these areas. These research areas include (1) quantification of variability and uncertainty in emission factors and emission inventories, including development of methods for dealing with small sample sizes, mixture distributions, censored data, dependencies between sampling distributions for parameters, inter-unit dependence, and autocorrelation, using a variety of techniques (Abdel-Aziz and Frey, 2003; Frey, 2003; Frey and Bammi 2002 and 2003; Frey and Bharvirkar, 2002; Frey, Bharvirkar, and Zheng, 1999; Frey and Li, 2003; Frey and Rhodes, 1996; Frey and Zheng, 2001, 2002a, 2002b; Zhao and Frey, 2003; Zheng and Frey, 2001); (2) quantification of uncertainty in the performance, emissions, and cost of advanced process technologies, such as coal-based gasification systems for production of power and chemicals, for the purpose of evaluating the potential pay-offs and downside risks of such technologies, comparison with conventional technologies, and identification of priorities to reduce uncertainty (Frey, 1998; Frey and Akunuri, 2001; Frey and Rubin, 1991a, 1991b, 1992a, 1992b, 1997; Frey, Rubin, and Diwekar, 1994; Frey and Tran, 1999); (3) optimization under uncertainty, including chance-constrained programming, stochastic optimization, and stochastic programming (Diwekar et al., 1997; Shih and Frey, 1995); (4) use of probabilistic methods as a means for gaining insight into needs for Federal involvement in research, development, and demonstration of energy technologies (Frey et al., 1995); (5) quantification of variability and uncertainty in human exposure and risk analysis, including development and recommendation of methods, development of software tools, and implementation of two dimensional probabilistic simulation methods as part of exposure and risk assessment models (Cullen and Frey, 1999; Frey and Burmaster, 1999; Frey and Rhodes, 1998; Frey and Rhodes, 1999; Frey, Zheng, Zhao, Li, and Zhu, 2002; Zheng and Frey, 2002a, 2002b, 2003); (6) evaluation of approximately a dozen sensitivity analysis methods with respect to applicability to food safety risk process models, including ability to deal with nonlinearity, thresholds, interactions, simultaneous variation in inputs, identification of factors contributing to high exposure outcomes, and other criteria and development of guidance for practitioners regarding the use of sensitivity analysis methods (Frey, Mokhtari, and Danish, 2003; Frey and Patil, 2002; Patil and Frey, 2003); and (7) development of requirements analysis for uncertainty and sensitivity analysis for a multimedia risk assessment framework (Loughlin et al., 2003). Sponsors of these activities have included the U.S. Department of Energy, U.S. Department of Agriculture, and the U.S. Environmental Protection Agency. Thus, there are clearly opportunities for these and other federal agencies to benefit from sharing information and developing or coordinating an integrated research agenda in the areas of uncertainty and sensitivity analysis. Although there have been a wide variety of applications and case studies, our research program has a common theme of developing, refining, or applying quantitative methods for uncertainty and sensitivity analysis, including the following considerations: (1) development of probability distributions for model inputs based upon statistical analysis of data or elicitation of expert judgment; (2) distinguishing between variability and uncertainty when appropriate to the assessment objective; (3) evaluation of alternative probability distributions models, parameter estimation methods, and goodness-of-fit techniques; (4) propagation of uncertainty typically using numerical methods but occasionally using analytical techniques; (5) evaluation of uncertainty in model outputs with respect to decision-making or risk management objectives, including identification of risks; (6) the use of sensitivity analysis methods to gain insights into key sources of uncertainty that should be priorities for additional data collection or research; and (7) the use of optimization methods under uncertainty to assist decision-making regarding technology design and environmental strategy development. The identification of research needs often is informed by working with realistic case studies. For example, in the process of quantifying uncertainty in hourly emissions from baseload coal-fired power plants for input to an air quality model, needs for dealing with inter-unit dependence and autocorrelation in the time series of emissions became apparent, thereby motivating a specific research program to address such needs (e.g., Abdel-Aziz and Frey, 2003). The prevalence of data containing non-detects, particular for air toxic emission factors but also in many other fields, motivates the need for development of methods for fitting distributions to censored data and estimating uncertainty in statistics estimated from such data (e.g., Zhao and Frey, 2003). Other examples include the need to develop methods for dealing with mixture distributions to more adequately represent variability in data such as for emission factors or exposure factors (e.g., Zheng and Frey, 2001). Thus, a key recommendation for developing research objectives that are policy relevant is to identify problems of medium to long term policy interest, identify methodological gaps that prevents a sufficiently thorough analysis and assessment, and target research to develop new methods to fill these gaps. Active areas of research and recommended areas for future investigation pertain to food safety risk assessment, $PM_{2.5}$ emissions and risk estimation, and development of integrated software tools to facilitate more widespread use of appropriate and rigorous methods for uncertainty and sensitivity analysis by practitioners, among others. Opportunities to learn across disciplines via workshops such as this should also be considered as a long-term interagency activity.

References

Abdel-Aziz, A., and H.C. Frey (2003), "Quantification of Hourly Variability in NOx Emissions for Baseload Coal-Fired Power Plants," *Journal of the Air & Waste Management Association*, submitted January 2003, accepted July 2003.

Cullen, A.C., and H.C. Frey (1999). *The Use of Probabilistic Techniques in Exposure Assessment: A Handbook for Dealing with Variability and Uncertainty in Models and Inputs.* Plenum: New York, 1999. 335 pages.

Diwekar, U.M., E.S. Rubin, and H.C. Frey (1997), "Optimal Design of Advanced Power Systems Under Uncertainty," *Energy Conversion and Management*, 38(15):1725–1735 (1997).

Frey, H.C. (1998), "Quantitative Analysis of Variability and Uncertainty in Energy and Environmental Systems," Chapter 23 in *Uncertainty Modeling and Analysis in Civil Engineering*, B. M. Ayyub, ed., CRC Press: Boca Raton, FL, 1998, pp. 381–423.

Frey, H.C. (2003), "Evaluation of an Approximate Analytical Procedure for Calculating Uncertainty in the Greenhouse Gas Version of the Multi-Scale Motor Vehicle and Equipment Emissions System ," Prepared for Office of Transportation and Air Quality, U.S. Environmental Protection Agency, Ann Arbor, MI, May 30, 2003.

Frey, H.C., and N. Akunuri (2001), "Probabilistic Modeling and Evaluation of the Performance, Emissions, and Cost of Texaco Gasifier-Based Integrated Gasification Combined Cycle Systems Using ASPEN," Prepared by North Carolina State University for Carnegie Mellon University and U.S. Department of Energy, Pittsburgh, PA, January 2001.

Frey, H.C., and S. Bammi (2002), "Quantification of Variability and Uncertainty in Lawn and Garden Equipment NOx and Total Hydrocarbon Emission Factors," *Journal of the Air & Waste Management Association*, 52(4):435–448 (April 2002).

Frey, H.C., and S. Bammi (2003), "Probabilistic Nonroad Mobile Source Emission Factors," *ASCE Journal of Environmental Engineering*, 129(2):162–168 (February 2003).

Frey, H.C., and R. Bharvirkar (2002), "Quantification of Variability and Uncertainty: A Case Study of Power Plant Hazardous Air Pollutant Emissions," Chapter 10 in *Human and Ecological Risk Analysis*, D. Paustenbach, Ed., John Wiley and Sons: New York, 2002. pp 587–617.

Frey, H.C., R. Bharvirkar, and J. Zheng (1999), Quantitative Analysis of Variability and Uncertainty in Emissions Estimation, Prepared by North Carolina State University for the U.S. Environmental Protection Agency, Research Triangle Park, NC. July 1999.

Frey, H.C., and D.E. Burmaster (1999), "Methods for Characterizing Variability and Uncertainty: Comparison of Bootstrap Simulation and Likelihood-Based Approaches," *Risk Analysis,* 19(1):109–130 (February 1999).

Frey, H.C., R.J. Lempert, G. Farnsworth, D.C. Acheson, P.S. Fischbeck, and E.S. Rubin (1995), A Method for Federal Energy Research Planning: Integrated Consideration of Technologies, Markets, and Uncertainties, Prepared by Carnegie Mellon, RAND, and Atlantic Council for Lawrence Livermore National Laboratory, Livermore, CA. April 1995.

Frey, H.C., and S. Li (2003), "Quantification of Variability and Uncertainty in AP-42 Emission Factors: Case Studies for Natural Gas-Fueled Engines," *Journal of the Air & Waste Management Association,* accepted for publication as of June 2003.

Frey, H.C., A. Mokhtari, and T. Danish (2002), "Evaluation of Selected Sensitivity Analysis Methods Based Upon Applications to Two Food Safety Risk Process Models," Draft, Prepared by North Carolina State University for Office of Risk Assessment and Cost-Benefit Analysis, U.S. Department of Agriculture, Washington, DC, December 2002.

Frey, H.C., and S.R. Patil (2002), "Identification and Review of Sensitivity Analysis Methods," *Risk Analysis,* 22(3):553–578 (June 2002).

Frey, H.C., and D.S. Rhodes (1996), "Characterizing, Simulating, and Analyzing Variability and Uncertainty: An Illustration of Methods Using an Air Toxics Emissions Example," *Human and Ecological Risk Assessment: an International Journal,* 2(4):762–797 (December 1996).

Frey, H.C., and D.S. Rhodes (1998), "Characterization and Simulation of Uncertain Frequency Distributions: Effects of Distribution Choice, Variability, Uncertainty, and Parameter Dependence," *Human and Ecological Risk Assessment: an International Journal,* 4(2):423–468 (April 1998).

Frey, H.C., and D.S. Rhodes (1999), Quantitative Analysis of Variability and Uncertainty in Environmental Data and Models: Volume 1. Theory and Methodology Based Upon Bootstrap Simulation, Report No. DOE/ER/30250, Vol. 1, Prepared by North Carolina State University for the U.S. Department of Energy, Germantown, MD, April 1999.

Frey, H.C., and E.S. Rubin (1991a), Development and Application of a Probabilistic Evaluation Method for Advanced Process Technologies, Final Report, DOE/MC/24248-3015, NTIS DE91002095, Prepared by Carnegie-Mellon University for the U.S. Department of Energy, Morgantown, West Virginia, April 1991, 364p.

Frey, H.C., and E.S. Rubin (1991b), "Probabilistic Evaluation of Advanced SO2/NOx Control Technology," *Journal of the Air and Waste Management Association,* 41(12):1585–1593 (December 1991).

Frey, H.C., and E.S. Rubin (1992a), "Evaluation of Advanced Coal Gasification Combined-Cycle Systems Under Uncertainty," *Industrial and Engineering Chemistry Research,* 31(5):1299–1307 (May 1992).

Frey, H.C., and E.S. Rubin (1992b), "Integration of Coal Utilization and Environmental Control in Integrated Gasification Combined Cycle Systems," *Environmental Science and Technology,* 26(10):1982-1990 (October 1992).

Frey, H.C., and E.S. Rubin (1997), "Uncertainty Evaluation in Capital Cost Projection," in *Encyclopedia of Chemical Processing and Design,* Vol. 59, J.J. McKetta, ed., Marcel Dekker: New York, 1997, pp. 480–494.

Frey, H.C., E.S. Rubin, and U.M. Diwekar (1994), "Modeling Uncertainties in Advanced Technologies: Application to a Coal Gasification System with Hot Gas Cleanup," *Energy* 19(4):449–463 (1994).

Frey, H.C., and L.K. Tran (1999), Quantitative Analysis of Variability and Uncertainty in Environmental Data and Models: Volume 2. Performance, Emissions, and Cost of Combustion-Based NOx Controls for Wall and Tangential Furnace Coal-Fired Power Plants, Report No. DOE/ER/30250, Vol. 2, Prepared by North Carolina State University for the U.S. Department of Energy, Germantown, MD, April 1999.

Frey, H.C., and J. Zheng (2001), Methods and Example Case Study for Analysis of Variability and Uncertainty in Emissions Estimation (AUVEE), Prepared by North Carolina State University for Office of Air Quality Planning and Standards, U.S. Environmental Protection Agency, Research Triangle Park, NC, February 2001.

Frey, H.C., and J. Zheng (2002a), "Quantification of Variability and Uncertainty in Utility NOx Emission Inventories," *J. of Air & Waste Manage. Association,* 52(9):1083–1095 (September 2002).

Frey, H.C., and J. Zheng (2002b), "Probabilistic Analysis of Driving Cycle-Based Highway Vehicle Emission Factors," *Environmental Science and Technology,* 36(23):5184–5191 (December 2002).

Frey, H.C., J. Zheng, Y. Zhao, S. Li, and Y. Zhu (2002), Technical Documentation of the AuvTool Software for Analysis of Variability and Uncertainty, Prepared by North Carolina State University for the Office of Research and Development, U.S. Environmental Protection Agency, Research Triangle Park, NC. February 2002.

Hanna, S.R., Z. Lu, H.C. Frey, N. Wheeler, J. Vukovich, S. Arunachalam, M. Fernau, and D.A. Hansen (2001), "Uncertainties in Predicted Ozone Concentrations due to Input Uncertainties for the UAM-V Photochemical Grid Model Applied to the July 1995 OTAG Domain," *Atmospheric Environment*, 35(5):891–903 (2001).

Loughlin, D., H.C. Frey, K. Hanisak, and A. Eyth (2003), "Implementation Requirements for the Development of a Sensitivity/Uncertainty Analysis Tool for MIMS," Draft, Prepared by Carolina Environmental Program and North Carolina State University for U.S. Environmental Protection Agency, Research Triangle Park, NC, May 6, 2003.

Patil, S.R., and H.C. Frey (2003), "Comparison of Sensitivity Analysis Methods Based Upon Applications to a Food Safety Risk Model," *Risk Analysis*, submitted December 19, 2002, accepted July 29, 2003.

Shih, J.S., and H.C. Frey (1995), "Coal Blending Optimization Under Uncertainty," European Journal of Operations Research, 83(3):452–465 (1995).

Zhao, Y., and H.C. Frey (2003), "Quantification of Uncertainty and Variability for Air Toxic Emission Factor Data Sets Containing Non-Detects," *Proceedings, Annual Meeting of the Air & Waste Management Association*, Pittsburgh, PA, June 2003.

Zheng, J., and H.C. Frey (2001), "Quantitative Analysis of Variability and Uncertainty in Emission Estimation: An Illustration of Methods Using Mixture Distributions," *Proceedings, Annual Meeting of the Air & Waste Management Association*, Pittsburgh, PA, June 2001.

Zheng, J., and H.C. Frey (2002a), AuvTool User's Guide, Prepared by North Carolina State University for the Office of Research and Development, U.S. Environmental Protection Agency, Research Triangle Park, NC. February 2002.

Zheng, J., and H.C. Frey (2002b), "Development of a Software Module for Statistical Analysis of Variability and Uncertainty," Proceedings, Annual Meeting of the Air & Waste Management Association, Pittsburgh, PA, June 2002

Zheng, J., and H.C. Frey (2003), "Windows-Based Software Implementation and Uncertainty Analysis of the EPA SHEDS/Pesticides Model," Proceedings, Annual Meeting of the Air & Waste Management Association, Pittsburgh, PA, June 2003.

Practical Strategies for Sensitivity Analysis Given Models with Large Parameter Sets

Terry Andres

University of Manitoba

Abstract

A model for sensitivity analysis purposes is a means of transforming from a set of input parameters to a single output value. Assume each input parameter has a domain of variability scaled to a uniform interval [0,1]. A model has a large parameter set if the number of parameters reaches the hundreds or thousands. Such models can arise through complex modeling projects, where many natural phenomena have an influence.

The cost of sensitivity analysis can grow as $O(N^2)$, where N is the number of parameters. The cost of running a single simulation (in man-hours, computer time, or dollars) can be $O(N)$, both in setup for a run, and in executing it. The number of runs required for sensitivity analysis can also be $O(N)$. If Resolution IV fractional factorial designs are used to estimate main effects of each parameter, at least $2N$ simulations are required for an initial estimate. A reduction in this rate of growth is highly desirable.

The problem simplifies because the number of influential parameters cannot grow as fast as the total number of parameters. Define influence of a parameter to be the fraction of variance of the model output that can be ascribed to that parameter. Only a small number of parameters (or interactions) can each contribute a significant fraction of the variance in the result. For instance, only 10 such effects (maximum) could contribute 10% or more of the final variance.

Some models may have *no* influential parameters. Then no amount of analysis would simplify an analyst's understanding of model behavior. An example is a model that takes the unweighted sum of a large number of input parameters (e.g., $f(X_1, X_2, ..., X_N) = \sum_{i=1}^{100} X_i$). None of the parameters can be considered to be influential on their own, as each contributes only 1% to the variance of the result. In such cases, sensitivity analysis may be inherently unrewarding.

Where sensitivity analysis is effective, one analyzes a model's performance to identify and characterize the influences of the small number of parameters. By analyzing different outcomes, intermediate results, and different transforms of outcomes, one might well identify a much larger number of parameters that have an identifiable influence, but for each specific analysis the goal is to find and study a small number of parameters.

If known, one could analyze that small group with a small design [e.g., a fractional factorial design with 16 parameters and 32 runs, in which each parameter takes extreme values (0 or 1)]. The iterated fractional factorial design approach [Andres and Hajas, 1993] suggested grouping all the parameters randomly into 16 groups instead, and then to vary all the parameters in a group alike. An analysis would indicate which group was most influential. After iterating the random grouping and application of a design many times, influential parameters are those that consistently appear in the most influential groups. The effectiveness of this approach depends on the fraction of the output variance that can be attributed to a small number of parameters. With a high signal-to-noise ratio, the number of runs required for an analysis varies as $O(\log N)$, rather than $O(N)$.

A simple fractional factorial design can only reveal main (linear) effects and two-parameter interactions. Andres [1997] showed how fractional factorial and Latin Hypercube Samples could be combined to give an unbiased mean, and to detect nonlinear influences. This approach was embedded in a tool called SAMPLE2 for generation of sample designs [Andres 1998]. The cost of this greater flexibility is a lower signal-to-noise ratio, meaning more iterations of an iterated experiment may be needed for the analysis.

Many practitioners feel that the simplest way of reducing the cost of sensitivity analysis is not to vary parameters that are thought to have little influence. This approach may not be justified when the models are implemented as computer programs in multidisciplinary projects. Beyond a certain level of complexity, no individual may completely understand the interplay of computed quantities. One of the chief benefits of sensitivity analysis is to determine what influences are driving a model so that specialists can assess model behaviour for plausibility. This outcome may not be achieved.

Nevertheless, it is possible to conduct sensitivity analysis with a small number of simulations, if the analyst has a good understanding of which parameters to study. It is important even in this case to repeat a few simulations with all parameters varying to check by paired analysis of variance that no significant source of variation has been overlooked.

References

Andres, T.H., and W.C. Hajas, 1993. "Using Iterated Fractional Factorial Design to Screen Parameters in Sensitivity Analysis of a Probabilistic Risk Assessment Model." Proceedings fot the Joint International Conference on Mathematical Methods and Supercomputing in Nuclear Applications, Karlsruhe, Germany, 1993 April 19–23, Vol. 2, pp. 328–37.

Andres, T.H. 1997. "Sampling Methods and Sensitivity Analysis for Large Parameter Sets." Journal of Statist. Comput. Simul., Vol. 57, pp. 77–110.

Andres, T.H. 1998. "SAMPLE2: Software to Generate Experimental Designs for Large Sensitivity Analysis Experiments." Proceedings of the Second International Symposium on Sensitivity Analysis of Model Output, Venice, Italy, Ca'Dolfin 1998 April 19–22.

An Integrated Regionalized Sensitivity Analysis and Tree-Structured Density Estimation Methodology

Olufemi Osidele and M. Bruce Beck

Environmental Informatics and Control Program, Warnell School of Forest Resources,
The University of Georgia, Athens, Georgia 30602

Regionalized Sensitivity Analysis (RSA) was developed in 1978 as a model-based technique for identifying critical uncertainties in current knowledge of environmental systems, and a basis for directing future research on such systems (Spear and Hornberger, 1980). RSA is founded on two principles—a qualitative definition of system behavior, and a binary classification of model simulations conditioned on the specified behavior definition. The behavior definition represents uncertainty about the external description of the system. It prescribes a set of constraints through which the model simulation trajectory must pass in order to qualify as an acceptable simulation of system behavior. The binary classification defines the model as exhibiting *behavior* (*B*) if the trajectory falls within the defined constraints, and *nonbehavior* (*NB*) if otherwise.

RSA employs Monte Carlo simulation and a Kolmogorov-Smirnov test to rank the uncertain model parameters according to their importance in discriminating between behavior and nonbehavior simulations. RSA is parameter-centric, in that it treats the parameters of the model as descriptors of the internal behavior of the system and indicators of the significance of their corresponding processes within the system. In other words, a sensitive parameter indicates a critical system process. Thus, RSA integrates uncertainties associated with both external and internal descriptions of the system. RSA has been applied in several model-based assessments, including parameter estimation for hydrological models (Hornberger *et al.*, 1985; Lence and Takyi, 1992), structural identification and hypothesis screening of ecological models (Osidele and Beck, 2001, 2003), and quality assurance of multimedia environmental models (Chen and Beck, 1999; Beck and Chen, 2000).

However, despite its ubiquity, RSA cannot identify multivariate correlation structures within the parameter space because the Kolmogorov-Smirnov test is conducted on marginal parameter distributions. This presents a problem for multimedia models which typically contain several interdependent parameters. For this reason, the results of RSA must be extended by multivariate statistical analyses, such as multiple regression and principal components analysis. Another such method is Tree-Structured Density Estimation (TSDE), a qualitative procedure for identifying parameter interactions (Spear, *et al.*, 1994). TSDE extends the concept of a histogram, into multidimensional space. It employs a sequence of recursive binary splits to partition the parameter domain into sub-domains comprising small regions of relatively high-density, and larger sparsely populated regions, similar, respectively, to the peaks and tails (or troughs) of a histogram. The result of the binary splitting is depicted as an inverted tree, where the root node represents the original parameter sampling domain, the other nodes are sub-domains, and the branches (the splits) are determined by most sensitive model parameters. The *terminal nodes* of the tree describe the final partitions of the original sampling domain.

When TSDE is applied to the behavior-producing parameter sets derived from RSA, tracing a high-density terminal node (HDTN) from the root node is equivalent to locating regions of the parameter space that have a high probability of matching the behavior definitions. Also, the sequence of parameters in the trace identifies the set of parameters that interact to produce a behavior simulation. Thus, the tree depicts, graphically and qualitatively, the multiple correlation structures that exist among the behavior-producing parameter values. Also, the combined volume of the HDTNs, in proportion to the overall sampling domain volume, indicates the probability of realizing the behavior definition. TSDE has been employed for comparative evaluation of stakeholder-derived environmental futures (Osidele, 2001; Beck, et al., 2002a).

The RSA–TSDE methodology incorporates the strengths of its component methods. Behavior definitions are composed for selected attributes of the system, and subsequently compared with the Monte Carlo simulation outputs. Figure 1 illustrates the RSA–TSDE methodology in a generalized framework for integrated systems assessment. The framework describes an adaptive approach for integrating the stakeholder and technical problems commonly associated with natural and built systems. Whereas the stakeholders are often most interested in the risks associated with the system (for example, potable water initially abstracted from a known polluted river or lake), the technical providers focus mainly on identifying priorities for advancing knowledge and better managing the performance of the system. Both stakeholder and technical problems are characterized by uncertainty. Examples include (i) the lack of consensus among stakeholders on their fears and desires for future environmental quality, and (ii) insufficient scientific knowledge to inform the design and operational management of environmental controls such as wastewater treatment plants and agricultural best management practices. These uncertainties are translated into numeric specifications and integrated with parameters and other decision variables in a model-based assessment. RSA identifies key individual parameters, which informs the prioritization of research, design, and systems management actions. TSDE identifies key groups of interdependent parameters, and estimates probabilities of meeting the prescribed specifications, which informs risk assessments. The feedback to stakeholders and technical providers renders the framework adaptive to changes in system policy and stakeholder concerns, as well as advancements in science and technology.

Adaptations of this framework have been applied to environmental systems problems, such as (i) stakeholder-science integration for generating environmental foresight (Osidele, 2001; Beck, et al., 2002b), and (ii) water quality management under the US EPA's Total Maximum Daily Load (TMDL) program (Osidele, et al., 2003). Presently, in a collaborative research program between the University of Georgia's Warnell School of Forest Resources and the EPA's Office of Research and Development, the RSA–TSDE methodology is being applied to uncertainty and sensitivity evaluation of the FRAMES-3MRA multimedia modeling and risk assessment system (Babendreier, 2003).

Keywords

Integrated Assessment, Modeling, Sensitivity, Simulation, Uncertainty.

References

Babendreier, J.E. (2003) National-scale multimedia risk assessment for hazardous waste disposal, *Proceedings, International Workshop on Uncertainty, Sensitivity and Parameter Estimation*, Federal, Interagency Steering Committee on Multimedia Environmental Modeling.

Beck, M.B., and J. Chen. (2000) Assuring the quality of models designed for predictive tasks. In A. Saltelli, K. Chan and M. Scott (eds.), *Sensitivity Analysis*, Wiley, Chichester.

Beck, M.B., J. Chen, and O.O. Osidele. (2002a) Random search and the reachability of target futures. In M.B. Beck (ed.), *Environmental Foresight and Models: a manifesto*, Elsevier, Oxford.

Beck, M.B., B.D. Fath, A.K. Parker, O.O. Osidele, G.M. Cowie, T.C. Rasmussen, B.C. Patten, B.G. Norton, A. Steinemann, S.R. Borrett, D. Cox, M.C. Mayhew, X-Q. Zeng, and W. Zeng. (2002b) Developing a concept of adaptive community learning: case study of a rapidly urbanizing watershed, *Integrated Assessment*, 3(4): 299–307.

Chen, J., and M.B. Beck. (1999) Quality assurance of multi-media model for predictive screening tasks, *Report EPA/600/R-98/106*, U.S. Environmental Protection Agency, Washington, DC.

Hornberger, G.M., K.J. Beven, B.J. Cosby, and D.E. Sappington. (1985) Shenandoah watershed study: calibration of a topography-based, variable contributing area hydrological model to a small forested catchment, *Water Resources Research*, 21(12): 1841–1850.

Lence, B.J., and A.K. Takyi. (1992) Data requirements for seasonal discharge programs: an application of a regionalized sensitivity analysis, *Water Resources Research*, 28(7): 1781–1789.

Osidele, O.O. (2001) Reachable Futures, Structural Change, and the Practical Credibility of Environmental Simulation Models, *Ph.D. Dissertation*, The University of Georgia, Athens, GA.

Osidele, O.O., and M.B. Beck. (2001) Identification of model structure for aquatic ecosystems using Regionalized Sensitivity Analysis, *Water Science and Technology*, 43(7): 271–278.

Osidele, O.O., and M.B. Beck. (2003) Food web modelling for investigating ecosystem behaviour in large reservoirs of the south-eastern United States: lessons from Lake Lanier, Georgia, *Ecological Modelling*, in press.

Osidele, O.O., W. Zeng, and M.B. Beck. (2003) Coping with uncertainty: a case study in sediment transport and nutrient load analysis, *Journal of Water Resources Planning and Management*, 129(4): 345–355.

Spear, R.C., and G.M. Hornberger. (1980) Eutrophication in Peel Inlet – II: Identification of critical uncertainties via generalized sensitivity analysis, *Water Research*, 14: 43–49.

Spear, R.C., T.M. Grieb, and N. Shang. (1994) Parameter uncertainty and interaction in complex environmental models, *Water Resources Research*, 30(11): 3159–3169.

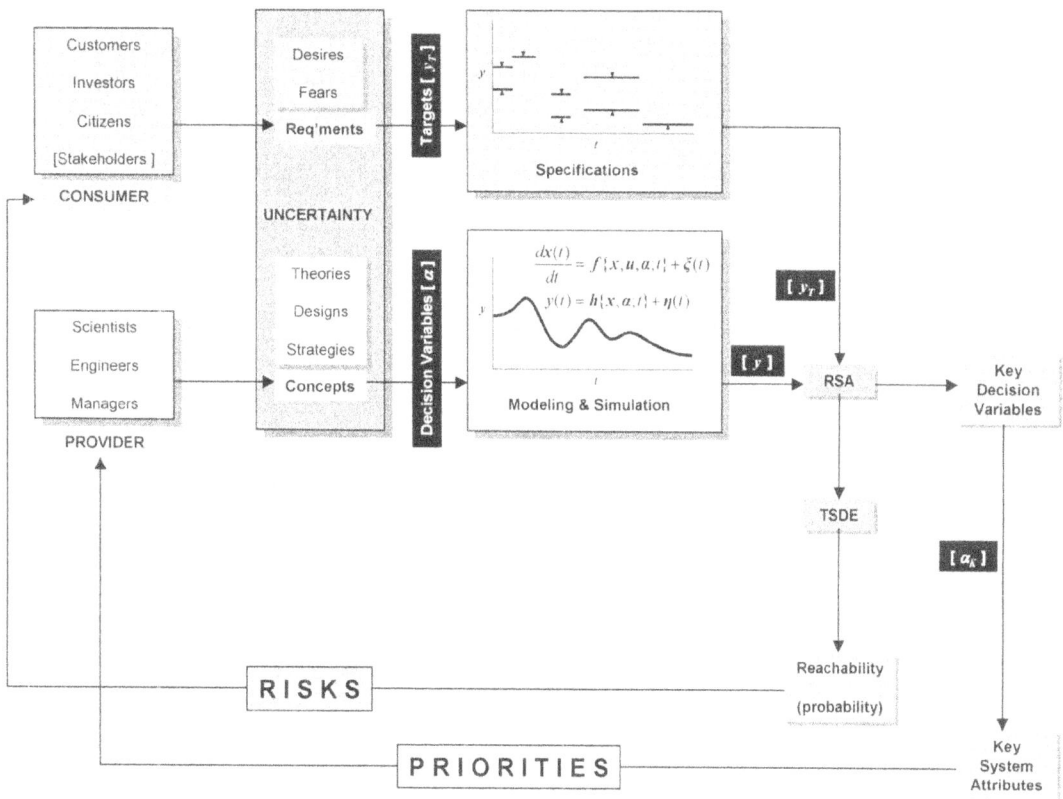

Figure 1: Generalized Framework for Integrated Systems Assessment

Sensitivity Analysis in the Context of Risk Significance

Sitakanta Mohanty

Center for Nuclear Waste Regulatory Analyses
Southwest Research Institute
6220 Culebra Road, San Antonio, Texas 78228

Email: smohanty@swri.org; Fax: (210) 522-5155

Introduction

The risk-informed, performance-based approach is increasingly being adopted by nuclear and non-nuclear industries (e.g., waste disposal, facility decommissioning, chemical process plant safety, and food safety) as a model for safety evaluation and licensing. Quantitative risk assessment, which permits systematic investigation, quantification, and explanation of the safety concept, is key to implementing the risk-informed, performance-based approach. The assessment is carried out probabilistically when a high degree of uncertainty is associated with the system. Sensitivity analysis (also referred to as uncertainty importance analysis in some contexts) is an important component of the probabilistic risk assessment (PRA) methodology. Results from sensitivity analysis typically are used to derive risk significance of various aspects of the system represented through parameters, conceptual models, and assumptions.

In the literature, parametric sensitivity analysis typically refers to the sensitivity of model outputs to various model parameters. Hundreds of parametric sensitivity analysis methods have been published in the literature (see Saltelli et al[1] for a recent review). The purpose of this presentation is to show how parametric and other sensitivity analyses results are used in determining risk significance. Rather than focusing on the details of various methods, this presentation highlights, through an example, some practical aspects and pitfalls in the traditional use of sensitivity analyses in determining risk significance. This presentation also highlights those approaches that can complement and overcome the limitations of traditional sensitivity analysis.

Work Description

The example PRA model[2] simulates a complex system characterized by numerous coupled processes, large heterogeneities, many length scales, long simulation periods, very slowly evolving processes, and very short duration-high consequence scenarios. This model has 330 sampled parameters (from a total of 950 parameters), 43 correlated parameters, 12 alternative conceptual models, and 6 primary subsystems or components. Various sensitivity analysis methods were applied to this example, including (i) parametric, (ii) distributional, (iii) conceptual model, and (iv) component sensitivity (e.g., what-if analysis) to identify the factors (i.e., parameters, distribution functions representing these parameters, alternative conceptual models, and subsystems, etc.) to which model output is sensitive.

Parametric sensitivity analysis used a number of different statistical and non-statistical methods[3-6]. Multiple methods were used in an effort to identify, as comprehensively as possible, those parameters that influence model outputs. Distributional sensitivity analysis[6] was conducted to identify the parameters for which the choice of the distribution function can significantly affect model output. Conceptual model sensitivity analysis[6] was used to identify potentially influential alternative models where the data are ambiguous. Component sensitivity analysis[6] was used to determine how degradation in the performance of major components influences model outputs.

Results and Conclusions

The use of a variety of parametric sensitivity analysis techniques resulted in the identification of a pool of influential parameters whose effects on risk merit further investigation. However, parametric sensitivity analysis alone did not always identify factors significant to risk, especially when the model output (i.e., the risk estimate) is far below the threshold of interest (e.g., the regulatory threshold or the design-basis threshold for product development) or when substantial level of conservatism is built into the model. For example, if the flow in the unsaturated zone is conservatively assumed to occur only in fractures, the attributes of the rock matrix will not be captured by parametric sensitivity analysis as important. In these cases, parametric sensitivity analysis is more helpful in establishing the correctness of the model than providing a compelling reason for reducing uncertainty in influential parameters. Equal attention should be given to understanding why the remainder of the parameters are not influential in parametric sensitivity analyses. The conceptual model and component sensitivity analyses used in this study appear to be useful for identifying those areas requiring further investigation from the perspective of gaining risk insights.

Finally, sensitivity analysis may not always provide a quantitative measure for ranking key factors or issues according to their risk significance. The linkage between the key factors or issues from sensitivity analysis results and the significance of those factors or issues to risk can be expressed only qualitatively (e.g., a factor/issue is of high, medium, or low importance to risk). Sensitivity analysis, however, can guide the analyst to probe the system model in an efficient and structured manner to answer how model assumptions, model limitations, data uncertainties, data inadequacy, data inaccuracy, subjectivity in data interpretation, and imprecision in results could influence risk significance.

Acknowledgments

The abstract was prepared to document work performed by the Center for Nuclear Waste Regulatory Analyses (CNWRA) for the U.S. Nuclear Regulatory Commission (NRC) under Contract No. NRC-02-97-009. The activities reported here were performed on behalf of the NRC's Office of Nuclear Material Safety and Safeguards (NMSS). The abstract is an independent product of the CNWRA and does not necessarily reflect the views or regulatory position of the NRC.

References

1. Saltelli, A., K. Chan, and E.M. Scott (Editors), Sensitivity Analysis, Wiley Series in Probability and Statistics, Chichester, New York: Wiley, 2000.

2. Mohanty, S., T.J. McCartin, D. Esh. "Total-System Performance Assessment Verison 4.0 Code: Module Description and User's Guide (Revised)," CNWRA Report, January 2002.

3. Lu, Y., and S. Mohanty. 2001. Sensitivity Analysis of a Complex, Proposed Geologic Waste Disposal System Using the Fourier Amplitude Sensitivity Test Method. Reliability Engineering and System Safety. Vol. 72 (3) pp 275–291.

4. Mohanty, S. and Y-T. (Justin) Wu. 2001. CDF Sensitivity Analysis Technique for Ranking Influential Parameters in the Performance Assessment of the Proposed High-Level Waste Repository at Yucca Mountain, Nevada, USA. Reliability Engineering and System Safety. Vol. 73 (2) pp. 167–176.

5. Mohanty, S. and J. Wu. 2002. Mean-based Sensitivity or Uncertainty Importance Measures for Identifying Influential Parameters. Probabilistic Safety Assessment and Management - PSAM6 (E.J. Bonano, A.L. Camp, M.J. Majors, R.A. Thompson, editors) Elsevier, New York, USA. Vol. 1 pp. 1079–1085.

6. Mohanty, S., R. Codell, J.M. Menchaca, R. Janetzke, M. Smith, P. LaPlante, M. Rahimi, A. Lozano. 2002. "System-Level Performance Assessment of the Proposed Repository at Yucca Mountain Using the TPA Version 4.1 Code," CNWRA 2002-05, Revision 1, San Antonio, TX: Center for Nuclear Waste Regulatory Analyses. (To be published).

5

SESSION 3:

UNCERTAINTY ANALYSIS APPROACHES, APPLICATIONS, AND LESSONS LEARNED — *IDENTIFICATION OF RESEARCH NEEDS*

Overview and Summary

Editor: Philip Meyer

The third session of the workshop comprised six presentations addressing uncertainty analysis methods. The case for uncertainty analysis/management in addressing complex environmental problems was succinctly made by one of the presenters, who noted that typical characteristics of these problems include high stakes decisions, disputed objectives and values, large uncertainties and knowledge gaps, the inability to delay decisions until the science is certain, and a reliance on models and assumptions. Although it is generally accepted that some evaluation of uncertainty is important, there is not yet a consensus on the appropriate methods to use in an uncertainty analysis. Most applications of uncertainty analysis to environmental modeling have been limited to an evaluation of the impact of uncertainty in model parameters. Methods to evaluate parameter uncertainty are well-established, if not always easy to implement in practice. Evaluation of prediction uncertainty using such methods implicitly assumes that model error is negligible. It has been observed by a number of authors, however, that model structural error may be much more significant than errors in model parameter values. The potential importance of model error and the need for better methods to evaluate it are becoming more widely recognized. Several of the presentations in this session discussed the development of methods to evaluate the impact of model error.

One of the presentations discussed model abstraction techniques applicable to multimedia environmental modeling. Model abstraction is relevant to generating parsimonious models while maintaining consistency with the available knowledge and data. Exploration of alternative models is often used to represent model uncertainty.

Uncertainty analysis in the environmental modeling arena has typically focused on quantitative methods applied to limited and well-defined targets, for example, the application of Monte Carlo simulation to derive an output distribution based on specified input distributions of a subset of model parameters. It has been increasingly recognized that this approach is often not satisfactory. Modeling complex, open environmental systems often results in significant, unquantifiable uncertainties remaining after the models have been formulated. In addition, it may be difficult to quantitatively account for all the (potentially conflicting) objectives of the various stakeholders involved in environmental management decisions. A challenging problem is the development of uncertainty analysis methods that appropriately consider subjective and non-quantitative factors. Several presentations in this session addressed this issue.

5.1.1 Discussion Summary

A number of relevant comments were made by participants during discussions. In many cases, more questions were raised during discussions than answered, indicating the need for additional research and development in this area.

There was general agreement that evaluation of model uncertainties must rely on observations. It is much easier to defend a model that has been tested against data than to defend either a model for which there is no evaluation data, or a model that cannot be tested because it does not predict a testable quantity. For multimedia environmental modeling applications, there may be no observations of the ultimate quantity predicted by the model (e.g., exposure to a contaminant). In this case, it may only be possible to calibrate a component of a multimedia model (e.g., the groundwater component). What impact does this have on the ability to estimate overall error and on the credibility of the multimedia model results?

The components of a multimedia environmental model may have widely varying credibility. How is it possible to account for the propagation of errors, particularly model structural errors, through such a system? Is it possible to have a credible multimedia model when one of its components is discounted by experts due, for example, to its over-simplification? Would the situation be improved by having all model components at a consistent level of credibility (measured how?), even if this meant that some knowledge and data were not ultimately used in the multimedia analysis?

The complexity of models should be driven by the purpose(s) the models are intended to fulfill. If this results in model simplification (abstraction), then it is important that the uncertainty associated with that simplification be assessed, particularly any introduced bias. It is unreasonable to expect a model to predict reality; a model may nonetheless prove useful. Model inadequacies need to be communicated openly to avoid misleading stakeholders and to assist decision makers.

5.1.2 Application Issues

The development of generic, easily implemented techniques and software to assess uncertainty in a comprehensive way was identified as an unmet need. Comprehensive uncertainty analysis includes consideration of uncertainty related to parameters, model structure, and forcing terms, and should be capable of representing quantitative and qualitative uncertainty concepts. Some early efforts at the development of such techniques were discussed, including extensions to the GLUE methodology, the NUSAP (Numeral Unit Spread Assessment Pedigree) method, a maximum likelihood Bayesian model averaging method, and modifications to the Regionalized Sensitivity Analysis method.

5.1.3 Research Needs

A variety of issues related to future research needs were identified by the presenters:

- Consideration of uncertainty deriving from the "social context" of the problem

- Methods of design for discovery of ignorance

- Improved understanding of models as evolving objects

- Guidance for evaluating very high-order, multimedia models under conditions of open model review by all stakeholders, a limited number of multi-disciplinary experts without conflict, and a sparse or non-existent history to match

- Improved representation of model structural uncertainty.

- Model performance measures that consider sources of error individually

- Generic techniques, relatively simple to implement, for model structural uncertainty assessment

- Consideration of the interaction between input errors and model structure

- Techniques that allow for the evaluation of model structures by hypothesis testing

- Methods to ensure that the space of potential model structures is adequately explored

- Dissemination into practice of state-of-the-art methods

- Using complementary uncertainty analysis techniques from various disciplines

- Improved understanding of the full spectrum of sources of uncertainty

- Improved understanding of the way uncertain knowledge can be used in the policy process

- Addressing institutional impediments to uncertainty management

5.1.4 Conclusions

Uncertainty analysis has historically emphasized assessment of the impact of parameter uncertainty. It is now recognized that this approach is inadequate. Developments in uncertainty analysis are currently centered on incorporating additional sources of uncertainty, most notably model structural uncertainty. The importance of including uncertain elements that can only be represented qualitatively is also gaining recognition. Significant impediments remain to the widespread application of comprehensive uncertainty analysis techniques.

Uncertainty: Foresight, Evaluation, and System Identification

M. Bruce Beck

Warnell School of Forest Resources, University of Georgia
Athens, Georgia 30602-2152

mbbeck@arches.uga.edu

The role of uncertainty and its analysis is addressed in three broad, inter-related domains, in exploring the future (foresight); coming to a judgement on whether the model is of a high or low quality (evaluation); and reconciling the behavior of the model with that apparently displayed in the past data (model calibration or, more broadly, system identification). Contemporary experience from each domain will be illustrated through a case study, before closing with some observations on possible requirements for future research.

Historical Trends

Let us first make some observations on the recent past, on how the subject of modeling, in respect to the analysis of uncertainty, has evolved over the past two decades or so. In this there has been a universal given, of course: the scope (order) of models generally expands (increases) with time. That said, our outlook on modeling, and the contexts in which models are applied for the purposes of assisting in the formulation of policy and the design of management actions, has been changing:

- From imagining we could identify constancy (and singularity) of structure in the behavior of the system ($f\{x,u,\alpha;t\}$),[1] i.e., invariance (and singularity) in the way in which the state variables of the model (x) are believed to be inter-connected (and previously, from imagining we could eventually identify the "truth" of the matter), to all this being an illusion;

- From imagining we could validate a model, in the conventional sense of (primarily) matching history, to data assimilation — wherein the data are merely assimilated into a prior model presumed to be entirely secure in its hypothetical basis, not employed ruthlessly to root out its inadequacies and inconsistencies;

- From supposing we could not rigorously address problems of system identification for data-sparse situations, to addressing "no-data" situations, including those, perhaps counter-intuitively, as imagined for the future;

- From being data-poor, to being data-rich yet information-poor, in the sense of being unable (satisfactorily) to reconcile high-order models (HOMs) with high-volume, high-quality sets of data — it was, after all, not difficult to reconcile a HOM with many parameters (α) against a sparse data set;

- From dealing exclusively with quantitative interpretations of uncertainty, gathered around the focus of the computational model, to broader interpretations having to do with who — besides the professional scientists developing the model — is affected by the composition of the model and its outcomes;

---

[1] For notational simplicity, and for clarity in the subsequent discussion, we assume here that the model expresses relationships (f), parameterized *via* α, among the system's inputs, u, state variables, x, and outputs, y (typically, observations of some of the state variables); t is time.

- From progressively and systematically excluding subjective, non-quantitative experience — of personal, non-instrumented observation of the system's behavior and our involvement with that behavior, including our personal imagination, hopes, and fears — to the near primacy of the scientifically lay stakeholder, including in matters of judgement relating to the quality of conventional, quantitative, computational models;

- From the stance of a command-and-control policy articulated through a privileged technocracy, to the democracy of a plurality of aspirations for the future.

To summarize, where once the analysis of uncertainty was seen as being confined to a computational model with a fixed structure, in which uncertainty was propagated into forecasts of the future from accounts of the uncertainty in the model's input disturbances, the initial conditions of its states, and its parameters (Beck, 1987), a broader picture now emerges. Reducing uncertainty by systematically increasing the scope of the model, with the expectation of ultimately converging on the discovery of an invariant, singular truth, is no longer presumed as the only prescription for modeling. We expect there to be structural change, especially over the increasing spans of our forecasting (and now observational) horizons, and structural uncertainty/error, possibly expressed as a plurality of candidate model structures populated by multitudes of candidate parameterizations.

The greatest changes, however, have not been in our perception of such sources of *scientific* uncertainty, but in our recognition of the sources of uncertainty entering into the broader picture from the *social context* in which modeling is carried out (as signaled, for example, in van Asselt and Rotmans (1996), as well as, more recently and more specifically, in Korfmacher (2001)). First, for instance, what is desired of the future, the target behavior of the model's outputs (y), may not be at all well defined. Its specification may emerge more through the democracy of "what the people want," than the technocracy of the singular, abstract decision-maker of the past. It may accordingly be a conflicting and self-contradictory plurality of widely ranged possibilities — nowhere near as clear and unambiguous as, for example, that the well concentration of benzene in water should always be less than a crisp, given, point value. Furthermore, what the people fear, and therefore wish to avoid in the future, may be as important as what they desire to have come about.

Second, numerical quantification of the model's inputs (u) may derive from scientifically lay stakeholders and be subject to the perceived reliability (uncertainty) with which policy actions (again u) are implemented. Think of the difference, in respect of the former, between the "population equivalent" inserted into the design calculations of a wastewater treatment plant and the self-reported practices of a farmer for quantifying the number of cattle grazing on a given pixel of pasture-land. Which of these rather different sorts of data would be regarded as the "more certain"? And for the latter, think of the uncertainty in the future performance of a Best Management Practice (BMP), such as a riparian buffer strip, relative to a wastewater treatment plant.[2] Better still, think of the operational reliability of a farmer adopting a particular fertilization or grazing pattern, *vis à vis* a professional wastewater treatment plant manager choosing to operate his/her plant for biological nutrient removal. Exactly how the activities of society are to be incorporated as quantitative inputs to the model are shifting, away from the (engineering) abstractions of the "population equivalent" and pump and valve control-settings, towards — in some instances — populating the models with agents acting as (scientifically lay) individuals with their own learning patterns and decision rules and their own cultural perspectives on man-environment interactions (Janssen and Carpenter (1999)).

Third, continuing this line of thought, the professional (scientific) expert and the (scientifically) lay stakeholder must accustom themselves to an intermingling of their "traditional" roles in the development and application of models. In the increasingly democratic processes of today, scientists and engineers have become a part of the problematique — no longer especially privileged as *primi*

---

[2] Recall that during planning it is generally assumed that, once constructed, a wastewater treatment facility will deliver a defined quality of effluent (the policy action u), with certainty, in principle, for all t.

inter pares (Beck *et al*, 2002b). In an age where science is to be "socially robust" (Gibbons, 1999; also Darier *et al*, 1999) their contributions will be subject to scrutiny, and the assurance of quality, not merely by their peers, but also by all who hold a stake in the issue for which that science is being purchased and produced. The bus driver, or garbage collector, or doctor, may legitimately comment on the veracity of the model ($f\{x,u,\alpha;t\}$) and its results (Funtowicz and Ravetz, 1993).

A qualitatively different dimension of uncertainty, with implications for the way in which we go about generating foresight, evaluating the quality of a model, and reconciling its behavior against the observations, must now at least be acknowledged. To a lesser or greater extent this will be apparent from the following experiences in some recent case studies.

Current Case Studies

Foresight: Foodweb Model of a Piedmont Impoundment

Here is a case that sought deliberately to respond — in the development and application of a model — to what the people hoped and feared for in the longer-term future, beyond the horizon of most policy-making for managing surface water quality (Beck *et al*, 2002b). Instead of defining the problem, acquiring the science, constructing the model, and making predictions, albeit qualified by uncertainty, we chose to work backward, as it were, from the aspirations of the community of stakeholders for the future, to their attainability viewed from the present. Our methods of analysis are reported in the companion presentation of Osidele and Beck on "An Integrated Regional Sensitivity Analysis and Tree-structured Density Estimation Methodology."

Imagine the gross uncertainty that must attach to the outcomes of a Foresight Workshop with some 30 or so lay stakeholders who, in an afternoon, must compose impressions of their worst fears and best hopes, a generation or two hence, for their cherished piece of the environment (in this instance, Lake Lanier in north Georgia). Our goals, broadly speaking, were (a) to assess the "reachability," or plausibility, of the community's hopes and fears and (b) to establish priorities for purchasing more science on those scientific unknowns — attaching to those parameters (α) — identified as key to either or both of the hopes and fears being realized. The essential question herein is, did such uncertainty about the target future domains of behavior (y), coupled with the equally large uncertainties attaching to a model of the lake's foodweb ($f\{x,u,\alpha;t\}$), which was somewhat speculative, render the analysis impotent in terms of attaining its goals?

The answer, was "no" (Beck *et al*, 2002b), essentially in line with results we have obtained from similar, but more conventional, studies of screening stormwater control strategies under uncertainty (Duchesne *et al*, 2001). It could have been otherwise. Either way, we would have been obliged to embark on another cycle of the analysis — indeed, in theory, an unending cycle of foresight generation — entirely in line with the goals of the procedure of the analysis, which we have called adaptive community learning (Beck *et al*, 2002b). We know what adaptive management is (Holling, 1978). In essence, policy therein (u) fulfils two functions: to probe the behavior of the environmental system in a manner designed to reduce uncertainty about that behavior, i.e., to enhance learning about the nature of the *physical* system ($f\{x,u,\alpha;t\}$); and to bring about some form of desired behavior (y) in that system. Adaptive community learning ought both to subsume the principles of adaptive management (so defined) and include actions, or a process of decision-making, whereby the community of stakeholders experiences learning about *itself*, its relationship with the valued piece of the environment, i.e., the community-environment relationship, and the functioning of the physical environment. Given the inter-generational prospect, the process will always be subject to gross uncertainty. Just as adaptive management celebrates a prudent measure of experimentation, so does adaptive community learning (Norton and Steinemann, 2001). The process will be one of "always learning, never getting it right" (Price and Thompson, 1997). In this, the community of stakeholders is interpreted in a much broader sense than merely stakeholders as policy persons/managers. Indeed, the scientifically lay stakeholder is pivotal in the procedure (Beck *et al*, 2002b).

Technically, the basic, seminal algorithm of Regionalized Sensitivity Analysis (RSA) (of Hornberger and Spear; see also the presentation of Osidele and Beck) had to be modified in two important respects: (i) the introduction of a multivariate, as opposed to a univariate, means of analysis for discriminating key from redundant model parameters (the Tree-Structured Density Estimation algorithm); and (ii) enhancement of the number of posterior, "behavior-giving" candidate parameterizations of the model, in order to increase the power of the preceding discrimination (a Uniform Coverage by Probabilistic Rejection sampling scheme). It also became apparent that the model in such analyses must be viewed as an evolving, fluid object. It is not something that would necessarily ever converge on a stable, fixed entity; a reasonably invariant software product, based on an essentially invariant science base, and with generic application potential, which our peer group of scientists and engineers might normally expect as an outcome. The model, in this context (if not more generally; Beck, 2002a), is more a vehicle designed, and continually redesigned, to explore a continually evolving problem space.

Evaluation: Predictive Exposure Model

When novel chemicals and other substances never previously encountered in the natural environment (supposedly, genetically modified materials) are to be manufactured, the call for a model to be employed in forecasting their fate and effects may be irresistible. But how should one evaluate — or validate, or assure the quality — of that model when, by definition, there is no history (y) to be matched? Is there anything more that could be done to buttress the conventional protocols of peer review, which essentially deal with approving or disapproving of the composition of $f\{x,u,\alpha;t\}$, cast essentially in the "internal" parametric space of the model, as opposed to judging the quality of the model in the complementary "external" space of its output performance (y), which is fundamental to the attribute of history having been matched?

Our case study was based on a precursor of one of the EPA's current multi-media models. It did not deal exactly with predicting the fate of an entirely novel chemical, if released into the environment, but with the migration of hazardous chemicals from storage facilities (Chen and Beck, 1999; Beck and Chen, 2000). Our method of analysis was the precursor, the basic RSA, of the methodology to be found in the presentation of Osidele and Beck. Given the uncertainty of a (presumed) complete absence of history, which may be closely approximated for some landfill sites, suppose we can specify the nature of the predictive task of the model — of projection into the entirely unknown — within the output space, y. For example, the goal of management might be to take action to avoid excessive levels of hazardous contaminants at nearby receptor sites, formally translated as two domains of target behavior — *not* observed history — to be matched by trajectories from the model.[3] We know that the RSA can discriminate between parameters within the vector α that are key and those that are redundant in determining whether the model can generate the target behaviors, or not. What is more, we know that the RSA was designed to do this under gross uncertainty, both in the composition of the model and in the specification of the output behavior, as outlined in the foregoing case study of the foodweb model. One can begin to conceive, therefore, of a candidate model that is well-suited (ill-suited) to its given, predictive task according to whether, say, its ratio of key/redundant parameters is high (low), for instance. This might be supplemented by considerations along the lines that a higher-quality model should have few so identified key parameters that are highly uncertain. In more common language, the test could deliver statements such as: the model is of high quality with respect to predicting high-end exposures but of poor quality for mean exposures. Alternatively, one could conceive of determining through this kind of test whether candidate model

---

[3] "Behavior," being concentrations above a critical maximum level, and "not-the-behavior," being its complement.

A is better suited to its predictive task than candidate model B, even for cases where the scopes of the models, the numbers of their parameters, differ significantly (Beck and Chen, 2000; Beck, 2002b). Importantly, however, judgements about the quality of the model are reflected back from the output space of the target behavior into attributes of the model's internal composition, i.e., the parameter vector (α) associated with the constituent hypotheses assembled into the model's structure.

A lesson was not so much learned from this case study. Rather, in wrestling with the issue of making methods of classical verification and validation relevant in what are essentially "no-data" contexts, a conceptual shift was achieved in both the ways in which a model can be viewed and hence the manner of its evaluation. Models, as Caswell observed long ago (Caswell, 1976), are objects designed to fulfil clearly expressed tasks, just as hammers, screwdrivers, and other tools have been designed to serve identified or stated purposes. Thus, the model may be used in the following variety of ways, some of which may seem unconventional, but in each of which its success as a *tool* must be evaluated (Beck, 2002b):

- As a succinctly encoded *archive* of contemporary knowledge about the behavior of a system;

- As an *instrument* of prediction;

- As a *device* for communicating scientific notions to a scientifically lay audience;

- As an exploratory *vehicle* for the discovery of ignorance.

This re-oriented perspective — of the model as a tool to be designed against a task specification — can be placed outside the traditional view of models as computerized articulations of theory whose purpose, at bottom, is to make predictions of a future state of nature, ultimately falsifiable by subsequent observation when the time comes.

At a technical level, we found, unsurprisingly, that no one number (or ratio of key/redundant parameters) could encapsulate sufficiently the notion of the "well-suitedness" of the model to its task (Beck and Chen, 2000). This might have to be captured within a frequency distribution of the numbers of constituent model parameters falling within each band of a continuum of the (Kolmogorov-Smirnov) test statistic for quantifying the significance of a parameter (from maximally key to maximally redundant).

System Identification: Models of Nutrient Cycles and Aquatic Microbial Systems

Our third case study is a largely familiar problem, differing from those of the past merely in that we have access to high-volume, high-quality data sets, in this instance, from a biological wastewater treatment system and an aquaculture pond. This is unusual in the study of water quality, in particular, where microbially mediated state interactions are significant. Absent hitherto adequate time-series data for the system's inputs (u) and outputs (y), it has become the rare exception for time-series models — the low-order models (LOMs) commonly referred to as "statistical" input-output or transfer-function models — to be encountered here, and respected. The prescription for reducing model uncertainty has been to press on toward an HOM, even a VHOM (very high-order model), for so runs the rhetoric: if everything of conceivable relevance has been included in the model, how can it possibly be "wrong"? To caricature the situation, there is customarily a stark imbalance between the order of the data (very low indeed) and the order of the models from the preferred modeling paradigm. We have HOMs that therefore cannot normally be shown unequivocally to be wrong — not as a whole, and certainly not at the level of our being able to accept some of the HOM's constituent hypotheses as adequate, while rejecting others as inadequate. And we have LOMs that can barely gain a foothold in any procedure for reconciling our theories with the empirical observations.

Access to high-volume, high-quality sets of data has the potential to transform this caricature of the situation. And so it does, but in ways that are not quite what might be expected. First, the HOMs fail, now demonstrably so against the better data, substantially so, indeed very substantially so, and even those considered industry standards. Second, let us be candid about the methods we all use. Old-fashioned, classical trial-and-error, without a formal hint of any uncertainty and without any "automated" search, i.e., entirely deterministic realizations of the HOM successively redirected by the analyst, seems the best means of securing the first 60–80% of a satisfactory interpretation of observed behavior. Third, in reconciling the model thus with the data, the issues are nearly always about eliminating structural error from the composition of $f\{x,u,\alpha;t\}$, notwithstanding any given prior HOM. It seems almost as if one must return to the beginning, to re-invent a better wheel. Fourth, the complementary LOMs, and their — by comparison — highly successful, automated identification from the data, fare well as univariate interpretations, but yield only hard-won insights into the nature of these biological systems, where so many of the state variables are interacting with each other in a dense, nonlinear, multivariable mesh.

As in the procedure of adaptive community learning emerging from the first case study, this third and last case study contributes to dealing with uncertainty primarily through a conceptual re-organization of existing methods, as here in rehabilitating the role of the LOM (a discussion of the accompanying prototype procedure of system identification can be found in Beck and Lin (2003)). No radically new methods for the analysis of uncertainty have emerged from this case study to serve better the procedure, except some extensions to the algorithms of recursive parameter estimation (based on Beck *et al*, 2002b). Pivotal in the procedure, however, is the notion that the parameters of a model should be presumed to be stochastic processes, essentially varying with time, not random variables (presumed constant but unknown). Symbolically, the procedure is underpinned by the outlook of $(\alpha(t))$, *not* α invariant with time; apparent, if not real, structural change is presumed to be present. In principle, $f\{x,u,\alpha;t\}$ will be changing through time, not merely in terms of how the state variables are believed to interact, but also in terms of the orders of state and parameter vectors (x,α) (Beck, 2002a).

Requirements for Future Research

On the matter of foresight, especially that intended as farsighted, addressing the issue of structural change, frequently under gross uncertainty, has been the subject of a recently published manifesto (Beck, 2002a). As the word "manifesto" suggests, this is a detailed and extensive statement of intentions, many of which are directed at the analysis of uncertainty, some of which have been realized in the first case study above, on adaptive community learning (Beck et al, 2002b).

Evaluation continues to be a critical area of continuing research, as evidenced by the establishment of the EPA's Council for Regulatory Environmental Models (CREM) and its ambitions with respect providing "Guidance on Environmental Models" (http://www.epa.gov/crem/sab). Ever larger models, i.e., VHOMs, will be constructed. They will be ever more dependent upon multi-disciplinary knowledge bases, extremely difficult to scrutinize, doubtless strongly immune to empirical refutation — as might be expected in the matching of history (witness our earlier brief remarks implying a trend toward data assimilation). Few peers may be available for their review, simply because there will be few peers for such VHOMs having no conflict of interest, as they are constructed and refined over ever longer project periods with ever larger project teams. There is therefore scope for much primary thought to be invested in the topic of evaluating VHOMs, in particular (Beck, 2002b). One might cast the issue, in the light of the above, as one of demonstrating that the model fulfills its designated task, without being unreasonably discordant with respect to whatever sparse history might be available (matching history, if it is of mildly dubious relevance to future behavior, plays

100

thus a secondary role). In this, at a technical level, the approach to examining the design of the model as such a tool should be further explored, using now the more comprehensive refinements of the RSA introduced in the companion presentation by Osidele and Beck.

With respect to system identification, other than the nascent procedural innovation touched upon in the third of the foregoing case studies, we have but one, specific, algorithmic recommendation for further research. Responding to the presumption of apparent structural change and error in the model of the system's behavior, the procedure would be better served if it were possible to incorporate a fixed interval smoothing (FIS) algorithm into current versions of the recursive prediction error algorithm for parameter estimation (see Beck *et al*, 2002a).

References

Beck, M.B. (1987), "Water Quality Modeling: A Review of the Analysis of Uncertainty," *Water Resources Research*, **23**, pp 1393–1442.

Beck, M.B. (ed) (2002a), "*Environmental Foresight and Models: A Manifesto,*" Elsevier, Oxford.

Beck, M.B. (2002b), "Model Evaluation and Performance," in *Encyclopedia of Environmetrics* (A.H. El-Sharaawi and W.W. Piegorsch, eds), pp 1275–1279.

Beck, M.B. and J. Chen, (2000), "Assuring the Quality of Models Designed for Predictive Purposes," in *Sensitivity Analysis* (A. Saltelli, K. Chan, and E.M. Scott, eds), Wiley, Chichester, pp 401–420.

Beck, M.B., and Z. Lin, (2003), "Transforming Data Into Information," *Water Science and Technology*, **47**(2), pp 43–51.

Beck, M.B., J.D. Stigter, and D. Lloyd Smith. (2002a), "Elasto-Plastic Deformation of the Structure," in *Environmental Foresight and Models: A Manifesto* (M B Beck, ed), Elsevier, Oxford, pp 323–350.

Beck, M.B., B.D. Fath, A.K. Parker, O.O. Osidele, G.M. Cowie, T.C. Rasmussen, B.C. Patten, B.G. Norton, A. Steinemann, S.R. Borrett, D. Cox, M.C. Mayhew, X-Q. Zeng, and W. Zeng. (2002b), "Developing a Concept of Adaptive Community Learning: Case Study of a Rapidly Urbanizing Watershed," *Integrated Assessment*, **3**(4), pp 299–307.

Caswell, H. (1976), "The Validation Problem", in *Systems Analysis and Simulation in Ecology*, Vol IV (B C Patten, ed), Academic, New York, pp 313–325.

Chen, J., and M.B. Beck. (1999), "Quality Assurance of Multi-media Model for Predictive Screening Tasks," *Report EPA/600/R-98/106,* Athens Environmental Research Laboratory, U.S. Environmental Protection Agency, Athens, Georgia.

Darier, E., C. Gough, B. De Marchi, S. Funtowicz, R. Grove-White, D. Kitchener, Â. Guimarães-Pereira, S. Shackley, and B. Wynne. (1999), "Between Democracy and Expertise? Citizens' Participation and Environmental Integrated Assessment in Venice (Italy) and St Helens (UK)," *J Environmental Policy & Planning*, **1**, pp 103–120.

Duchesne, S., M.B. Beck, and A.L.L. Reda. (2001), "Ranking Stormwater Control Strategies Under Uncertainty: The River Cam Case Study," *Water Science and Technology*, **43**, pp 311–320.

Funtowicz, S.O., and J.R. Ravetz. (1993), "Science for the Post Normal Age," *Futures*, **25**(7), pp 739–755.

Gibbons, M (1999), "Science's New Social Contract with Society," *Nature*, **402** (Supplement), pp C81–C84 (2 December).

Holling, C.S. (ed) (1978), "*Adaptive Environmental Assessment and Management,*" Wiley, Chichester.

Janssen, M.A., and S.R. Carpenter. (1999), "Managing the Resilience of Lakes: A Multi-agent Modeling Approach," *Conservation Ecology*, **3**(2), 15 [online].

Korfmacher, K.S. (2001), "The Politics of Participation in Watershed Modeling," *Environmental Management*, **27**(2), pp 161–176.

Norton, B.G., and A. Steinemann. (2001), "Environmental Values and Adaptive Management," *Environmental Values*, **10**(4), pp 473–506.

Price, M.F., and M. Thompson. (1997), "The Complex Life: Human Land Uses in Mountain Ecosystems," *Global Ecology and Biogeography Letters*, **6**, pp 77–90.

Van Asselt, M.B.A., and J.Rotmans. (1996), "Uncertainty in Perspective," *Global Environmental Change*, **6**(2), pp 121–157.

Uncertainty in Environmental Modelling: A Manifesto for the Equifinality Thesis

Keith Beven

University of Lancaster, United Kingdom

The Equifinality Thesis

In a series of papers from Beven (1993–2003) on, I have made the case and examined the causes for an approach to hydrological modelling based on a concept of equifinality of models and parameter sets in providing acceptable fits to observational data. The Generalised Likelihood Uncertainty Estimation (GLUE) methodology of Beven and Binley (1992), developed out of the Hornberger-Spear-Young (HSY) method of sensitivity analysis (Hornberger and Spear, 1981), has provided a means of model evaluation and uncertainty estimation from this perspective (see Beven et al., 2000; Beven and Freer, 2001; Beven, 2001, for summaries of this approach). In part, the origins of this concept lie in purely empirical studies that have found many models giving good fits to data.

There is a very important issue of modelling philosophy involved that might explain some of the reluctance to accept the thesis. Science, including hydrological science, at the macroscales at which we are interested in making predictions for the sensible management of resources, is supposed to be an attempt to work toward a single correct description of reality. It is not supposed to conclude that there must be multiple feasible descriptions of reality. The users of research also do not (yet) expect such a conclusion and might then interpret the resulting ambiguity of predictions as a failure (or at least an undermining) of the science. This issue has been addressed directly by Beven (2002a), who shows that equifinality of representations is not incompatible with a scientific research program, including formal hypothesis testing. In that paper, the modelling problem is presented as a mapping of the landscape into a space of feasible models (structures as well as parameter sets, see also Beven, 2002b). At least for deterministic model runs, the uncertainty does not lie in the predictions within this model space. The uncertainty lies in how to map the real system into that space of feasible models. Mapping to an "optimal" model is equivalent to mapping to a single point in the model space. Statistical evaluation of the covariance structure of parameters around that optimal model is equivalent to mapping to a small contiguous region of the model space. Mapping of Pareto optimal models is equivalent to mapping to a front or surface in the space of performance measures but which might be a complex manifold with breaks and discontinuities when mapped into in the model space. But computer-intensive studies of responses across the model space have shown that these mappings are too simplistic, since they arbitrarily exclude many models that are very nearly as good as the "optima." For any reasonably complex model, good fits are commonly found much more widely than just in the region of the "optimum" or Pareto "optima" (quotation marks are used here because the apparent global optimum may change significantly with changes in calibration data, errors in input data or performance measure).

Equifinality and Deconstructing Model Error

This also brings attention to the problem of model evaluation and the representation of model error. The GLUE methodology has been commonly criticised from a statistical inference viewpoint for using subjective likelihood measures and not using a formal representation of model error (e.g., Clarke, 1994; Thiemann et al., 2001; and many different referees). For ideal cases, this can mean that non-minimum error variance (or maximum likelihood) solutions might be accepted as good models, that the resulting likelihoods do not provide the true probabilities of predicting an output given the model, while the parameter estimates might be biased by not taking the correct structural model of the errors into account in the likelihood measure. In fact, the GLUE methodology is general in that it can use "formally correct" likelihood measures if this seems appropriate (see Romanowicz et al., 1994, 1996; and comments by Beven and Young, 2003), but need not require that any single model is correct (and "correct" here normally means not looking too closely at some of the assumptions made about the real errors in formulating the likelihood function, even if, in principle, those assumptions can be validated e.g., the assumption that model structure can be treated as "true" and the error treated as an additive "measurement error").

Another View of Model Evaluation

So what are the implications of taking an alternative view, one in which it is accepted that the hydrological model (and the error model) may not be structurally correct and that there may not be a clear optimal model, even when multiple performance measures are considered? This situation is not rare in hydrological modelling. It is commonplace. It should, indeed, be *expected* because of the overparameterisation of hydrological models, particularly distributed models, relative to the observational data available for calibration (even in research catchments). But modellers rarely search for good models that are not "optimal." Nor do they often search for reduced dimensionality models that would provide equally good predictions, but which might be more robustly estimated (e.g. Young, 2001, 2002; Beven and Young, 2003). Nor do they often consider the case where the "optimal" model is not really acceptable (see, for example, Freer et al., 2002); it is, after all, the best available.

Perhaps the problems stem from the continuing idea that model errors can be treated as additive (or multiplicative) "measurement errors" with the consequent (often implicit) assumption that the model is in some sense correct. This may be acceptable in the search for an optimal model, but not necessarily acceptable if we are really searching for models that are behavioural in the sense of being acceptably consistent with the available data. The model evaluation process can then be inverted by searching for all the potential models that are within the range of observation error. However, any model evaluation of this type needs to take account of the multiple sources of model error more explicitly (Beven and Young, 2003). This is difficult for realistic (rather than idealised) cases. Simplifying the sources of error to input errors, model structural errors and true measurement errors is not sufficient because, of the potential for incommensurability between observed and predicted variables (most modellers simply assume that they are the same quantity, even where this is clearly not the case). Thus, in assessing model acceptability, it is really necessary to decide on an appropriate level of "*effective observation error*" that takes account of such differences. When defined in this way, the effective observation error need not have zero mean or constant variance, nor need it be Gaussian in nature, particularly where there may be physical constraints on the nature of that error. Once this as been done, then it should be required that any behavioural model should provide *all* its predictions within the range of this effective observational error.

Equifinality and Assessing Predictive Uncertainty

For those models that meet such a criterion and are then retained as behavioral, it would be possible to use a weight, based on past performance, in using the predictions to assess the uncertainty in potential outcomes (in a way similar to the current GLUE methodology). This methodology gives rise to some interesting possibilities. If a model does not provide predictions within the specified range, then it should be rejected as non-behavioural. Within this framework, there is no possibility of a representation of model error being allowed to compensate for poor model performance, even for the "optimal" model. If there is no model that proves to be behavioral, then it is an indication that there are conceptual, structural, or data errors (although it may still be difficult to decide which is the most important). There is, perhaps, more possibility of learning from the modelling process on occasions when it proves necessary to reject all the models tried.

This implies that consideration also has to be given to input and boundary condition errors, since, as noted before, even the "perfect" model might not provide behavioural predictions if it is driven with poor input data error. Thus, it should be the combination of input/boundary data realisation (within reasonable bounds) and model parameter set that should be evaluated against the observational error. Any compensation effect between an input realisation and model parameter set in achieving success in the calibration period will then be implicitly included in the set of behavioural models.

This approach will be discussed in the context of an application to rainfall-runoff modelling in the presentation.

References

Beven, K.J., Prophecy, reality and uncertainty in distributed hydrological modelling, *Adv. Water Resourc.*, 16, 41–51, 1993.

Beven, K.J., *Rainfall-Runoff Modelling: The Primer*, Wiley, Chichester, 2001.

Beven, K.J., Towards a coherent philosophy for Environmental Modelling, *Proc. Roy. Soc. Lond.*, 2002a.

Beven, K.J., Towards an alternative blueprint for a physically-based digitally simulated hydrologic response modelling system, *Hydrol. Process.*, 16, 2002b.

Beven, K.J., and A.M. Binley, The future of distributed models: model calibration and uncertainty prediction, *Hydrological Processes*, 6, 279–298, 1992.

Beven, K.J., J. Freer, B. Hankin, and K. Schulz, 2000, The use of generalised likelihood measures for uncertainty estimation in high order models of environmental systems. in Nonlinear and Nonstationary Signal Processing, W.J. Fitzgerald, R.L. Smith, A.T. Walden, and P.C. Young (Eds). CUP, 115–151.

Beven, K.J., and J. Freer, 2001 Equifinality, data assimilation, and uncertainty estimation in mechanistic modelling of complex environmental systems, *J. Hydrology*, 249, 11–29.

Beven, K.J., and P.C. Young, 2003, Comment on Bayesian Recursive Parameter Estimation for Hydrologic Models, by M. Thiemannm, M. Trosset, H. Gupta, and S. Sorooshian, *Water Resources Research* 39(5), DOI: 10.1029/2001WR001183

Clarke, R.T., 1994 *Statistical Modelling in Hydrology*, Wiley: Chichester.

Freer, J. E., K.J. Beven, and N.E. Peters. 2002, Multivariate seasonal period model rejection within the generalised likelihood uncertainty estimation procedure. in *Calibration of Watershed Models*, edited by Q. Duan, H. Gupta, S. Sorooshian, A.N. Rousseau, and R. Turcotte, AGU Books, Washington, 69–87.

Hornberger, G.M., and R.C. Spear. 1981, An Approach to the Preliminary Analysis of Environmental Systems, *J. Environmental Management*, 12, 7–18.

Romanowicz, R., K.J. Beven and J. Tawn, 1998, Evaluation of Predictive Uncertainty in Non-Linear Hydrological Models Using a Bayesian Approach, in V. Barnett and K.F. Turkman (Eds.) *Statistics for the Environment II . Water-Related Issues*, Wiley, 297–317.

Romanowicz, R., K.J. Beven and J. Tawn, 1996, Bayesian Calibration of Flood Inundation Models, in M.G. Anderson, D.E.Walling and P. D. Bates, (Eds.) *Floodplain Processes*, 333–360.

Thiemann, M, M. Trosset, H. Gupta,and S. Sorooshian, 2001, Bayesian Recursive Parameter Estimation for Hydrologic Models, *Water Resourc. Res.*

Young, P C., Data-Based Mechanistic Modelling and Validation of Rainfall-Flow Processes, in Anderson, M G and Bates, P D (Eds), *Model Validation: Perspectives in Hydrological Science*, Wiley, Chichester, 117–161, 2001.

Young, P C., 2002, Advances in Real-Time Flood Forecasting, *Philosophical Transactions of the Royal Society: Mathematical, Physical and Engineering Sciences*, A360, 1433–1450,.

Model Abstraction Techniques
Related to Parameter Estimation and Uncertainty

Yakov Pachepsky[1], Rien van Genuchten[2], Ralph Cady[3], and Thomas J. Nicholson[3]

[1]Environmental Microbial Safety Laboratory, USDA-ARS, ypachepsky@anri.barc.usda.gov;
[2]U.S. Salinity Laboratory, USDA-ARS, rwang@ussl.ars.usda.gov;
[3]U.S. Nuclear Regulatory Commission, Office of Nuclear Regulatory Research,
TJN@nrc.gov and REC2@nrc.gov

Model abstraction is a methodology for reducing the complexity of a simulation model while maintaining the validity of the simulation results with respect to the question that the simulation is being used to address (Frantz, 2002). The need for model abstraction has been recognised in simulations of complex engineering and military systems that show that increased level of detail does not necessarily imply increased accuracy of simulation results, but usually increases computational complexity and may make simulation results more difficult to interpret. Similar observations have been made for simulations of subsurface flow and transport problems. Model abstractions that lead to reduced computational overhead and complexity can enable risk assessments to be run and analyzed with much quicker turnaround, with the potential for allowing further analyses of problem sensitivity and uncertainty. In addition, because of the highly heterogeneous nature of the subsurface, the issues of data collection and parameter estimation are as essential as computational complexity. While increased levels of detail in the data currently do not necessarily imply increased accuracy of the simulations, it usually does imply increased data collection density. Finally, model abstraction is important in enhancing communication. Simplifications that result from appropriate model abstractions may make the description of the problem more easily relayed to and understandable by others, including decision-makers and the public. It is often imperative to explicitly acknowledge the abstraction strategy used and its inherent biases, so that the modeling process is transparent and tractable.

Model abstraction explicitly deals with uncertainties in model structure. Model abstraction techniques and examples of their application in subsurface flow and transport include (a) using pre-defined hierarchies of models, (b) simplifying process descriptions based on the specific range of input parameters, i.e., reducing dimensionality, (c) parameter elimination based on simulation results, i.e., sensitivity analysis, (d) combining system states whose distinctions are irrelevant to the simulation output, i.e., combining individual stream tubes in a stochastic transport model, or upscaling based on aggregation, (e) dividing a model into loosely connected components, executing each component separately, and searching for constraints that execution of one component can impose on other components, i.e., running a flow model independently of the transport model, (f) combining states involving similar sequences and distinctions among the individual sequences that are irrelevant to the final outcome, i.e., abstracting the iterative plume construction to the transport of particle ensembles undergoing non-Brownian motion, (g) replacing continuous variables by class variables, i.e., using regression trees to develop pedotransfer functions used for hydraulic parameters estimations, or genetic algorithms in model calibration optimization, (h) temporal aggregation, i.e., replacing several closely-spaced events with a single event, (i) aggregating entities in a natural hierarchical structures, i.e., replacing a heterogeneous soil profile with an equivalent homogenous profile, (j) function aggregation to provide a coarser list of states or output information from existing entities, i.e., representing the water regime of a soil layer by means of either infiltration or evaporation, while neglecting redistribution, (k) using probabilistic inputs to develop lumped models, i.e., statistical averaging of flow and transport behavior for temporal and spatial upscaling, (l) using look-up tables to simplify the input-output transformation within a model or model

107

component by means of a decrease in computational effort, (m) rule-based solutions of model equations, i.e., using cellular automata in flow and transport problems, (n) metamodeling with neural networks, i.e., neural network approximations of a range of output scenarios for a particular remediation site, (o) spatial correlation-based metamodeling, i.e., using spatial correlations in flow and transport data assimilation, and (p) wavelet-based metamodeling.

Applications of model abstraction require criteria to select a simpler model, justify validity, and quantify questions being addressed. The criteria have yet to be developed based on quantified uncertainty and cost-benefit analyses. For purposes of vadose zone water flow and solute transport modeling, simplicity may be related to the number of processes being considered explicitly in the simulations, details of the discretization, runtime, number of measurements for parameter estimation, and correlations among parameters. Validity must be related to variability in data and to the uncertainty in the simulation results. Questions being addressed relate to specific outputs defined in terms of probability thresholds or physical thresholds for pre-defined locations in space in time.

During the first phase of this project we developed prospective directions for testing the model abstraction process using high-density data sets for water flow in typical environments. We concluded that, for model abstraction in flow and transport model *development*, the prospective direction should be on *model structure* modifications, whereas the prospective direction for model abstraction in flow and transport model *parameterization* should be on *model behavior* modification. Similarly, the prospective direction for model abstraction in flow and transport *simulations* should be on *model form modification*. Field data sets for a humid environment (Fig. 1) and for an arid environment (Fig. 2) were selected based on their completeness and complexity to explore specific issues, e.g., complex three-dimensional processes rendered as two and one-dimensional processes, or replacing directly measured soil hydraulic properties by pedotransfer function estimates. Future work with field data sets will compare the efficiency of model analysis techniques and provide a basis for developing rule-based strategies for model abstraction in the area of subsurface water and solute transport.

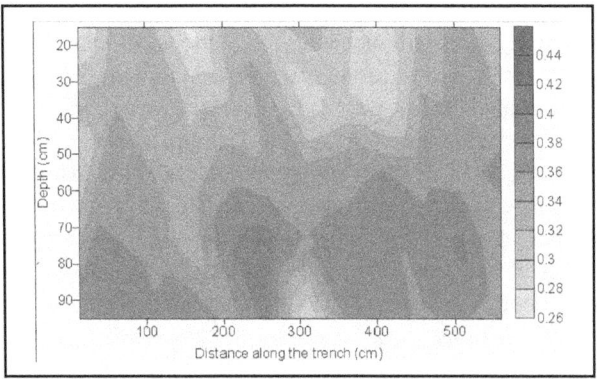

Figure 1. A snapshot of soil water contents monitored for 1 year (along with pressure heads and solute concentrations) along a trench in a loamy soil at the Bekkevoort site, Belgium.

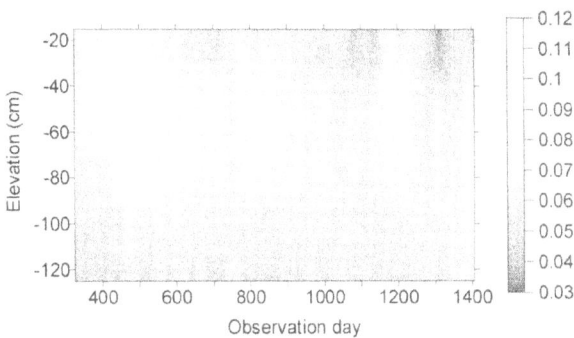

Figure 2. Image map of soil water contents monitored for 3 years (along with soil temperatures) at the USGS Amargosa Research Site, Nevada.

References

Diederik, J., J. Simunek, A.Timmerman and J.Feyen, 2002, "Calibration of Richards' and convection–dispersion equations to field-scale water flow and solute transport under rainfall conditions," *Journal of Hydrology*, 259: 15–31. (Belgium field study)

Frantz, Frederick K., "A Taxonomy of Model Abstraction Techniques," Computer Sciences Corporation, One MONY Plaza, Mail Drop 37-2, Syracuse, NY, June 2003. (Available at the U.S. Air Force Research Laboratory's Web site: http://www.rl.af mil/tech/papers/ModSim/ModAb.html)

Pachepsky, Yakov, Martinus Th. van Genuchten, Ralph Cady, and Thomas J. Nicholson, "Letter Report: Task 1— Identification and Review of Model-Abstraction Techniques," U.S. Department of Agriculture, Agricultural Research Service, Beltsville, Maryland, February 27, 2003.

USGS, Amargosa Desert Research Site Web site: http://nv.usgs.gov/adrs/, 2003.

Toward a Synthesis of Qualitative and Quantitative Uncertainty Assessment: Applications of the Numeral, Unit, Spread, Assessment, Pedigree (NUSAP) System

Jeroen van der Sluijs[a], Penny Kloprogge[a], James Risbey[b], and Jerry Ravetz[c]

[a] Copernicus Institute for Sustainable Development and Innovation, Department of Science Technology and Society, Utrecht University, The Netherlands (j.p.vandersluijs@chem.uu.nl).
[b] School of Mathematical Sciences, Monash University, Clayton, Australia
[c] Research Method Consultancy (RMC), London

Abstract: A novel approach to uncertainty assessment, known as the NUSAP method (Numeral Unit Spread Assessment Pedigree) has been applied to assess qualitative and quantitative uncertainties in three case studies with increasing complexity: (1) the monitoring of VOC emissions from paint in the Netherlands, (2) the TIMER energy model, and (3) two environmental indicators from the Netherlands 5th Environmental Outlook. The VOC monitoring involves a simple calculation scheme with 14 parameters. The TIMER model is a complex non-linear dynamic system model, which consists of over 300 parameters. The indicators in the Environmental Outlook result from calculations with a whole chain of soft-linked model calculations, involving both simple and complex models. We show that the NUSAP method is applicable not only to simple but also to complex models in a meaningful way and that it is useful to assess not only parameter uncertainty but also (model) assumptions. The method provides a means to prioritize uncertainties and focus research efforts on the potentially most problematic parameters and assumptions, identifying at the same time specific weaknesses in the knowledge base. With NUSAP, nuances of meaning about quantities can be conveyed concisely and clearly, to a degree that is quite impossible with statistic methods only.

Keywords: uncertainty; pedigree; NUSAP; quality; environmental assessment; assumption ladenness

Introduction

In the field of environmental modeling and assessment, uncertainty studies have mainly involved quantitative uncertainty analysis of parameter uncertainty. These quantitative techniques provide only a partial insight into what is a very complex mass of uncertainties. In a number of projects, we have implemented and demonstrated a novel, more comprehensive approach to uncertainty assessment, known as the NUSAP method (acronym for Numeral Unit Spread Assessment Pedigree). This paper presents and discusses some of our experiences with the application of the NUSAP method, using three case studies with increasing complexity.

NUSAP and the Diagnostic Diagram

NUSAP is a notational system proposed by Funtowicz and Ravetz (1990), which aims to provide an analysis and diagnosis of uncertainty in science for policy. It captures both quantitative and qualitative dimensions of uncertainty and enables one to display these in a standardized and self-explanatory way. The basic idea is to qualify quantities using the five qualifiers of the NUSAP acronym: Numeral, Unit, Spread, Assessment, and Pedigree. By adding expert judgment of reliability (Assessment) and systematic multi-criteria evaluation of the production process of numbers (Pedigree), NUSAP has extended the statistical approach to uncertainty (inexactness) with the methodological (unreliability) and epistemological (ignorance) dimensions.

111

NUSAP acts as a heuristic for good practice in science for policy by promoting reflection on the various dimensions of uncertainty and making these explicit. It provides a diagnostic tool for assessing the robustness of a given knowledge base for policymaking and promotes criticism by clients and users of all sorts—expert and lay—and will thereby support extended peer review processes.

NUSAP yields insights on two independent properties related to uncertainty in numbers, namely spread and strength. Spread expresses inexactness, whereas strength expresses the quality of the underlying knowledge base, in view of its methodological and epistemological limitations. The two metrics can be combined in a diagnostic diagram mapping strength and sensitivity of model parameters. The diagnostic diagram is based on the notion that neither spread alone nor strength alone is a sufficient measure for quality. Robustness of model output to parameter strength could be good even if parameter strength is low, provided that the model outcome is not critically influenced by the spread in that parameter. In this situation, our ignorance of the true value of the parameter has no immediate consequences because it has a negligible effect on model outputs. Alternatively, model outputs can be robust against parameter spread even if its relative contribution to the total spread in model is high, provided that parameter strength is also high. In the latter case, the uncertainty in the model outcome adequately reflects the inherent irreducible uncertainty in the system represented by the model. Uncertainty then is a property of the modeled system and does not stem from imperfect knowledge on that system. Mapping model parameters in a diagnostic diagram thus reveals the weakest critical links in the knowledge base of the model with respect to the model outcome assessed, and helps in the setting of priorities for model improvement.

Case I: A Simple Model

Emissions of VOCs (Volatile Organic Compounds) from paint in the Netherlands are monitored in the framework of VOC emission reduction policies. The annual emission figure is calculated from a number of inputs: national sales statistics of paint for five different sectors, drafted by an umbrella organization of paint producers; paint import statistics from Statistics Netherlands (lump sum for all imported paint, not differentiated to different paint types); an assumption on the average VOC percentage in imported paint; an assumption on how imported paint is distributed over the five sectors; and expert guesses for paint-related thinner use during application of the paint.

We developed and used a NUSAP-based protocol for the assessment of uncertainty and strength in emission data (Risbey *et al.*, 2001), which builds *inter alia* on the Stanford Protocol (Spetzler and von Holstein, 1975) for expert elicitation of probability density functions to represent quantifiable uncertainty and extends it with a procedure to review and elicit parameter strength, using a pedigree matrix. The expert elicitation systematically makes explicit and utilizes unwritten insights in the heads of experts on the uncertainty in emission data, focusing on limitations, strengths, and weaknesses of the available knowledge base.

Pedigree conveys an evaluative account of the production process of information, and indicates different aspects of the underpinning of the numbers and scientific status of the knowledge used. Pedigree is expressed by means of a set of pedigree criteria to assess these different aspects. The pedigree criteria used in this case are proxy, empirical basis, methodological rigor, and validation. Assessment of pedigree involves qualitative expert judgment. To minimize arbitrariness and subjectivity in measuring strength, a pedigree matrix is used to code qualitative expert judgments for each criterion into a discrete numeral scale from 0 (weak) to 4 (strong) with linguistic descriptions (modes) of each level on the scale. Table 1 presents the pedigree matrix we used in this case study.

Code	Proxy	Empirical	Method	*Validation*
4	Exact measure	Large sample direct measurements	Best available practice	Compared with independent measurements of same variable
3	Good fit or measure	Small sample direct measurements	Reliable method, commonly accepted	Compared with independent measurements of closely related variables
2	Well correlated	Modeled/ derived data	Acceptable method, limited consensus on reliability	Compared with measurements not independent
1	Weak correlation	Educated guesses/ rule-of-thumb estimate	Preliminary methods, unknown reliability	Weak/indirect validation
0	Not clearly related	Crude speculation	No discernible rigor	No validation

Table 1. Pedigree matrix for emission monitoring. Note that the columns are independent.

The expert elicitation interviews start with an introduction of the task of encoding uncertainty and a discussion of pitfalls and biases associated with expert elicitation (such as motivational bias overconfidence, representativeness, anchoring, bounded rationality, lamp-posting, and implicit assumptions).

	Proxy	*Empirical*	*Method*	*Validation*	*Strength**
NS-SHI	3	3.5	4	0	0.7
NS-B&S	3	3.5	4	0	0.7
NS-DIY	2.5	3.5	4	3	0.8
NS-CAR	3	3.5	4	3	0.8
NS-IND	3	3.5	4	0.5	0.7
Th%-SHI	2	1	2	0	0.3
Th%-B&S	2	1	2	0	0.3
Th%-DIY	1	1	2	0	0.25
Th%-CAR	2	1	2	0	0.3
Th%-IND	2	1	2	0	0.3
Imported paint	3	4	4	2	0.8
VOC % imp.	1	2	1.5	0	0.3

Table 2. Pedigree scores for input parameters.
*The *Strength* column averages and normalizes the scores on a scale from 0 to 1.

Note: NS=National Sales, Th%=Thinner use during application of paint (SHI, B&S, DIY, CAR, and IND refer to each of the five sectors)

Next, the expert is asked to indicate strengths and weaknesses in the knowledge base available for each parameter. This starts with an open discussion and then moves to the pedigree criteria that are discussed one by one for each parameter, ending with a score for each criterion (Table 2).

The protocol is designed to stimulate creative thinking on conceivable sources of error and bias. We identified 5 disputable basic assumptions in the monitoring calculation, and 15 sources of error and 4 conceivable sources of motivational bias in the data production.

In a next step in the interview, the expert is asked to quantify the uncertainty in each parameter as a PDF using a simplified version of the Stanford protocol (see *Risbey et al.*, 2001 for details). We used the PDFs elicited as input for a Monte Carlo analysis to assess propagation of parameter uncertainty and the relative contribution of uncertainty in each parameter to the overall uncertainty in VOC emission from paint. We found that a range of ±15% around the average for total 1998 VOC emission from paint (52 ktonne) captures 95% of the calculated distribution.

We further analyzed the uncertainty using a NUSAP diagnostic diagram (Fig. 1) to combine results from the sensitivity analysis (relative contribution to variance, Y-axis) and pedigree (strength, X-axis). Note that the strength axis is inverted, left-hand corresponds to a strong and right-hand to a weak knowledge base.

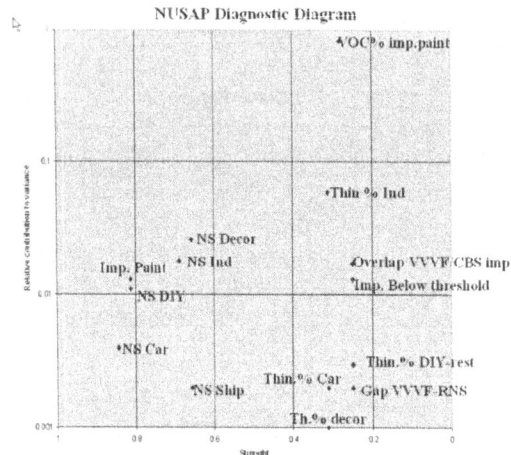

Figure 1 *Diagnostic diagram for VOC from paint*

The diagnostic diagram identified uncertainty regarding the assumed VOC percentage of imported paint as the most problematic. Other input quantities in the VOC monitoring calculations whose uncertainty was diagnosed to be "important" are assumed percentage of additional thinner use for paint applied in industry, the overlap between the paint import statistics and the national paint sales statistics, and import in volumes below the import statistics reporting threshold. The case is documented in detail in Van der Sluijs *et al.* (2002a).

Case II: A Complex Model

The TIMER (Targets IMage Energy Regional model) model is part of RIVM's Integrated Model to Assess the Global Environment (IMAGE). TIMER is an energy model that, amongst others, was used in the development of the 2001 greenhouse gas emission scenarios from the Inter-Governmental Panel on Climate Change (IPCC). We used the so-called B1 scenario produced with IMAGE/TIMER for the IPCC Special Report on Emissions Scenarios as case study.

Using the Morris (1991) method for global sensitivity analysis we explored quantitative uncertainty in parameters in terms of their relative importance in influencing model results. TIMER is a non-linear model containing a large number of input variables. The Morris method is a sophisticated algorithm where parameters are varied one step at a time in such a way that if sensitivity of one parameter is contingent on the values that other parameters may take, the Morris method is likely to capture such dependencies. TIMER contains 300 variables. Parameters were varied over a range from 0.5 to 1.5 times the default values. The method and full results are documented in Van der Sluijs *et al.* (2002b).

114

The analysis clearly differentiated between sensitive and less sensitive model components. Also, sensitivity to uncertainty in a large number of parameters turned out to be contingent on the particular combinations of samplings for other parameters, reflecting the non-linear nature of several parts of the TIMER model. The following input variables and model components were identified as most sensitive with regard to model output (projected CO_2 emissions):

- Population levels and economic activity;

- Variables related to the formulation of intra-sectoral structural change of an economy;

- Progress ratios to simulate technological improvements, used throughout the model;

- Variables related to resources of fossil fuels (size and cost supply curves);

- Variables related to autonomous and price-induced energy efficiency improvement;

- Variables related to initial costs and depletion of renewables;

We assessed parameter pedigree by means of a NUSAP expert elicitation workshop. 19 experts on the fields of energy economy and energy systems analysis and uncertainty assessment attended the workshop. We limited the elicitation to those parameters identified either as sensitive by the Morris analysis or as a "key uncertain parameter" in an interview with one of the modelers. Our selection of variables to address in the NUSAP workshop counted 39 parameters. To further simplify the task of reviewing parameter pedigree, we grouped together similar parameters for which pedigree scores might be to some extent similar. This resulted in 18 clusters of parameters. For each cluster a pedigree-scoring card was made, providing definitions and elaborations on the parameters and associated concepts, and a scoring part to fill out the pedigree scores for each parameter. We used the same criteria and pedigree matrix as in the VOC case (table 1), but added a fifth criterion: *theoretical understanding*. This is because the theoretical understanding of the dynamics of the energy system is in its early stage of development. The modes for this pedigree criterion are: Well-established theory (4); accepted theory partial in nature (3); partial theory limited consensus on reliability (2); preliminary theory (1); and crude speculation (0).

For the expert elicitation session, we divided the participants into three parallel groups. Each participant received a set with all 18 cards. Assessment of parameter strength was done by discussing each of the parameters (one card at a time) in a moderated group discussion addressing strengths and weaknesses in the underpinning of each parameter, focusing on, but not restricted to, the five pedigree criteria. Further, we asked participants to provide a characterization of value-ladenness. A parameter is said to be value-laden when its estimate is influenced by ones preferences, perspectives, optimism, or pessimism or co-determined by political or strategic considerations. Participants were asked to draft their pedigree assessment as an *individual* expert judgment, informed by the group discussion.

We used radar diagrams, and kite diagrams (Risbey *et al.*, 2001) to graphically represent results (Fig. 2). Both representations use polygons with one axis for each criterion, having 0 in the center of the polygon and 4 on each corner point of the polygon. In the radar diagrams, a line connecting the scores represents the scoring of each expert. The kite diagrams follow a traffic light analogy. The minimum scores in each group for each pedigree criterion span the green kite; the maximum scores span the amber kite. The remaining area is red. The width of the amber band represents expert disagreement on the pedigree scores. In some cases the size of the green area was strongly influenced by a single deviating low score given by one of the experts. In those cases the light green kite shows what the green kite would look like if that outlier had been omitted. A kite diagram captures the information from all experts in the group without the need to average expert opinion.

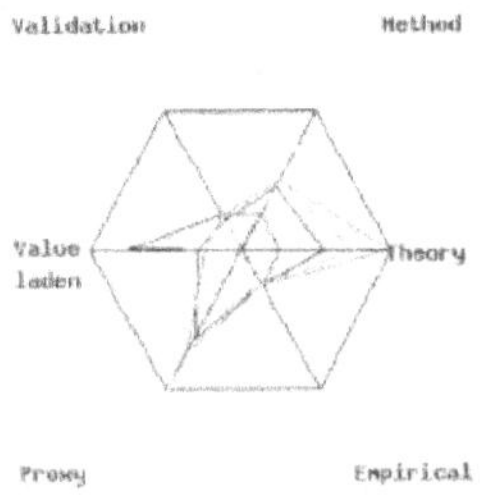

Figure 2a. Example of radar diagram of the gas depletion multiplier assessed by six experts.

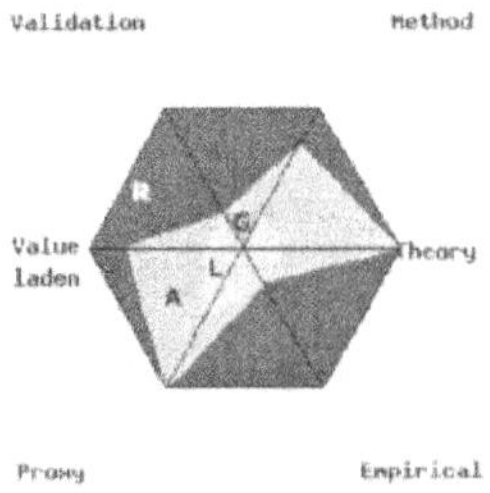

Figure 2b. same, but represented as kite diagram. G=green, L=light green, A=amber, R=red

Results from the sensitivity analysis and strength assessments were combined in Figure 3 to produce a diagnostic diagram.

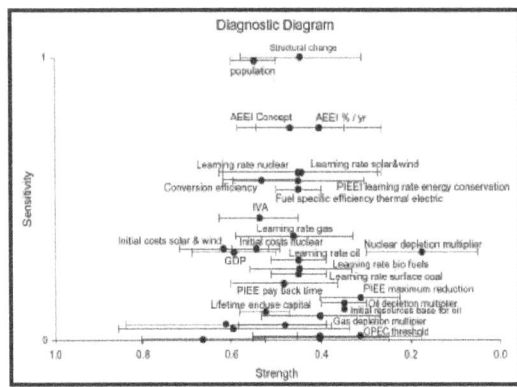

Figure 3. Diagnostic diagram for key uncertainties in TIMER model parameters.

The diagram shows each of the reviewed parameters plotted. The sensitivity axis measures (normalized) importance of quantitative parameter uncertainty. The strength axis displays the normalized average pedigree scores. Error bars indicate one standard deviation about the average expert value, to reflect expert disagreement on pedigree scores. The strength axis has 1 at the origin and zero on the right. In this way, the more "dangerous" variables are in the top right quadrant of the plot (high sensitivity, low strength).

We identified three parameters as being close to the danger zone: Structural change, B1 population scenario, and Autonomous Energy Efficiency Improvement (AEEI). These variables have a large bearing on the CO_2 emission result, but have only weak to moderate strength as judged from the pedigree exercise.

When variables are particularly low in strength, the theory, data, and method underlying their representation may be weak and we can then expect that they are less perfectly represented in the model. With such high uncertainty on their representation, it cannot be excluded that a better representation would give rise to a higher sensitivity. An example of such a variable could be the nuclear depletion multiplier, which has a strength from almost none to weak and a moderate sensitivity contribution.

116

Case III: Chains of Models

As input for the Netherlands Environmental Policy Plan, the Netherlands Environmental Assessment Agency (EAA/RIVM) prepares every 4 years an assessment of key environmental indicators outlining different future scenarios for a time period of 30 years: the National Environmental Outlook (EO). It presents hundreds of indicators reflecting the pressure on or state of the Dutch, European, or global environment. Model calculations play a key role in the assessments. In a "model chain" of soft-linked computer models—varying in complexity—effects regarding climate, nature, and biodiversity, health and safety, and the living environment are calculated for different scenarios. The total of model and other calculations and operations can be seen as a "calculation chain." Often, these chains behind indicators involve many analysts from several departments within the RIVM. Many assumptions have to be made in combining research results in these calculation chains, especially since the output of one computer model often does not fit the requirements of input for the next model (scales, aggregation levels).

We developed a NUSAP-based method to systematically identify, prioritize and analyze importance and strength of assumptions in these model chains including potential value-ladenness. We demonstrated and tested the method on two EO5 indicators: "change in length of the growth season" and "deaths and emergency hospital admittances due to tropospheric ozone."

We identified implicit and explicit assumptions in the calculation chain by systematic mapping and deconstruction of the calculation chain, based on document analysis, interviews and critical review. The resulting list of key assumptions was reviewed and completed in a workshop. Ideally, importance of assumptions should be assessed based on a sensitivity analysis. However, a full sensitivity analysis was not attainable because varying assumptions is much more complicated than, for instance, changing a parameter value over a range; it often requires construction of a new model. Instead, we used the expert elicitation workshop not only to review pedigree of assumptions but also to estimate their quantitative importance.

Score	2	1	0
Plausibility	plausible	acceptable	fictive or speculative
Inter-subjectivity peers	many would make same assumption	several would make same assumption	few would make same assumption
Inter-subjectivity stakeholders	many would make same assumption	several would make same assumption	few would make same assumption
Choice space	hardly any alternative assumptions available	limited choice from alternative assumptions	ample choice from alternative assumptions
Influence situational limitations (time, money, etc.)	choice assumption hardly influenced	choice assumption moderately influenced	totally different assumption when no limitations
Sensitivity to view and interests of the analyst	choice assumption hardly sensitive	choice assumption moderately sensitive	choice assumption sensitive
Influence on results	only local influence	greatly determines the results of link in chain	greatly determines the results of the indicator

Table 3. Pedigree matrix for reviewing the knowledge base of assumptions

Table 3 presents the pedigree matrix used in this study. In the workshop, the experts indicated on scoring cards (one card for each assumption) how they judge the assumption on the pedigree criteria and how much influence they think the assumption has on results. An essential part of our method

117

is that a moderated group-discussion takes place in which arguments for high or low scores per criterion are exchanged and discussed. In this way experts in the group remedy each other's blind spots, which enriches the quality of the individual expert judgments. We deliberately did not ask a consensus judgment of the group, because we consider expert disagreement a relevant dimension of uncertainty.

Assumptions that have a low score on both influence on the results and on the pedigree criteria can be qualified as "weak links" in the chain of which the user of the assessment results needs to be particularly aware.

Analysis of the calculation chain of the indicator "change in length of the growth season" yielded a list of 23 assumptions. The workshop participants selected seven assumptions as being the most important ones. These were reviewed using the pedigree matrix and prioritized according to estimated influence. Combining the results, the weakest links (high influence, low strength) in the calculation chain turned out to be the choice for a GCM (General Circulation Model, projecting time series of geographic patterns of temperature change as a function of greenhouse forcing) and the assumption that the scenarios used for economic development were suitable for the EO5 analyses for the Netherlands and that the choice for the range in global greenhouse gas emission scenarios used was suitable for the global analysis.

Analysis of the calculation chain of the indicator 'deaths and hospital admittances due to exposure to ozone' yielded a list of 24 assumptions. 14 key-assumptions were selected by the workshop participants as the most important ones, and prioritized. Combining the results of pedigree analysis and estimated influence, the following assumptions showed up as the weakest links of the calculation chain: Assumption that uncertainty in the indicator is only determined by the uncertainty in the Relative Risk (RR is the probability of developing a disease in an exposed group relative to those of a non-exposed group as a function of ozone exposure) and the assumption that the global background concentration of ozone is constant over the 30 year time horizon. The full EO5 case and method for the review of assumptions is documented in Kloprogge *et al.* (2003).

Conclusion

We have implemented and demonstrated the NUSAP method to assess qualitative and quantitative uncertainties in three case studies with increasing complexity: a simple model, a complex model, and environmental indicators stemming from calculations with a chain of models.

The cases have shown that the NUSAP method is applicable not only to simple but also to complex models in a meaningful way and that it is useful to assess not only parameter uncertainty but also (model) assumptions. A diagnostic diagram synthesizes results of quantitative analysis of parameter sensitivity and qualitative review (pedigree analysis) of parameter strength. It provides a useful means to prioritize uncertainties according to quantitative and qualitative insights.

The task of quality control in complex models is a complicated one and the NUSAP method disciplines and supports this process by facilitating and structuring a creative process and in depth review of qualitative and quantitative dimensions of uncertainty. It helps to focus research efforts on the potentially most problematic parameters and assumptions, identifying at the same time specific weaknesses in the knowledge base.

Similar to a patient information leaflet alerting the patient to risks and unsuitable uses of a medicine, NUSAP enables the delivery of policy-relevant quantitative information together with the essential warnings on its limitations and pitfalls. It thereby promotes the responsible and effective use of the information in policy processes. With NUSAP, nuances of meaning about quantities can be conveyed concisely and clearly, to a degree that is quite impossible with statistic methods only.

References

Funtowicz, S.O., and J.R. Ravetz, *Uncertainty and Quality in Science for Policy*. Kluwer, 229 pp., Dordrecht, 1990.

Kloprogge, P, J.P. van der Sluijs, and A. Petersen, *A method for the analysis of assumptions in assessments applied to two indicators in the fifth Dutch Environmental Outlook*, Department of Science Technology and Society, Utrecht University, 2004.

Morris, M.D., Factorial sampling plans for preliminary computational experiments, *Technometrics*, Vol. 33, Issue 2, 1991.

Risbey, J.S., J.P. van der Sluijs and J. Ravetz, *Protocol for Assessment of Uncertainty and Strength of Emission Data*, Department of Science Technology and Society, Utrecht University, report nr. E-2001-10, 22 pp, Utrecht, 2001 (www.nusap.net).

C.S. Spetzler, and S. von Holstein, Probability Encoding in Decision Analysis, *Management Science*, 22(3), (1975).

Van der Sluijs, J.P., J. Risbey, and J. Ravetz, *Uncertainty Assessment of VOC emissions from Paint in the Netherlands*, Department of Science Technology and Society, Utrecht University, 2002a, 90 pp (www.nusap.net).

Van der Sluijs, J.P., J. Potting, J. Risbey, D. van Vuuren, B. de Vries, A. Beusen, P. Heuberger, S. Corral Quintana, S. Funtowicz, P. Kloprogge, D. Nuijten, A. Petersen, J. Ravetz., *Uncertainty assessment of the IMAGE/TIMER B1 CO$_2$ emissions scenario, using the NUSAP method* Dutch National Research Program on Climate Change, Report no: 410 200 104, 227 pp, Bilthoven, 2002b, 237 pp (www.nusap.net).

Conceptual and Parameter Uncertainty Assessment via Maximum Likelihood Bayesian Model Averaging

Shlomo P. Neuman[1], Ming Ye[2], and Philip D. Meyer[2]

[1]University of Arizona, Neuman@hwr.arizona.edu
[2]Pacific Northwest National Laboratory, Ming.Ye@pnl.gov and Philip.Meyer@pnl.gov

Hydrologic analyses typically rely on a single conceptual-mathematical model. Yet hydrologic environments are open and complex, rendering them prone to multiple interpretations and mathematical descriptions. Adopting only one of these may lead to statistical bias and underestimation of uncertainty. Bayesian Model Averaging (BMA) (*Hoeting et al.*, 1999) provides an optimal way to combine the predictions of several competing conceptual-mathematical models and to assess their joint predictive uncertainty. *Neuman and Wierenga* (2003) have recently developed a comprehensive strategy for constructing alternative conceptual-mathematical models of subsurface flow and transport, selecting the best among them, and using them jointly to render optimum predictions under uncertainty. A key element of this strategy is a Maximum Likelihood (ML) implementation of BMA (MLBMA) proposed by *Neuman* (2002, 2003). It renders BMA computationally feasible by basing it on a ML approximation of model posterior probability due to *Kashyap* (1982) and the ML parameter estimation methods of *Carrera and Neuman* (1986a) for deterministic models and *Hernandez et al.* (2002, 2003) for stochastic moment models. The approach incorporates both site characterization and site monitoring data so as to base the outcome on an optimum combination of prior information (scientific and site knowledge plus data) and model predictions.

We apply MLBMA to geostatistical models of log air permeability data obtained from single-hole pneumatic injection tests in six vertical and inclined boreholes drilled into unsaturated fractured tuff at the Apache Leap Research Site (ALRS) in central Arizona. Seven alternative omni-directional variogram models of log permeability are postulated for the site: power (characteristic of a random fractal), exponential without or with first- or second-order polynomial drift, and spherical with similar drift options. The data do not support accounting for directional effects by considering the variograms to be anisotropic. Unbiased ML estimates of variogram parameters and drift coefficients are obtained using Adjoint State Maximum Likelihood Cross Validation (ASMLCV) (*Samper and Neuman*, 1989a) in conjunction with Universal Kriging (UK) and Generalized Least Squares (GLS). Commonly used information criteria (AIC, BIC, and KIC) provide an ambiguous ranking of the models, which does not justify selecting one of them and discarding the rest as is commonly done in practice. Instead, we eliminate three of the models based on their negligibly small ML-based posterior probability and use the remaining four models, with the corresponding ML variogram parameter and drift coefficient estimates, to project the measured log permeabilities by kriging onto a rock volume that includes but extends beyond the six test boreholes. We then average these four projections, and associated kriging error variances, using the posterior probability of each model as weight. Figure 2 depicts the resulting MLBMA log permeability projections and associated error variances across a vertical cut through the volume. Finally, we cross-validate the results by eliminating from consideration all data from one borehole at a time, repeating the above process, and comparing the predictive capability of MLBMA with that of each individual model. The comparison entails performing conditional Monte Carlo simulations of log permeability throughout the volume using each model, evaluating the corresponding cumulative distribution functions, and averaging them across all models using their posterior probabilities as weights. We find that (Table 1) MLBMA combines a relatively low predictive log score (small amount of lost information) with high predictive coverage (large proportion of predictions falling within the MC generated 90% prediction interval), rendering it superior to any individual geostatistical model of log permeability at the ALRS.

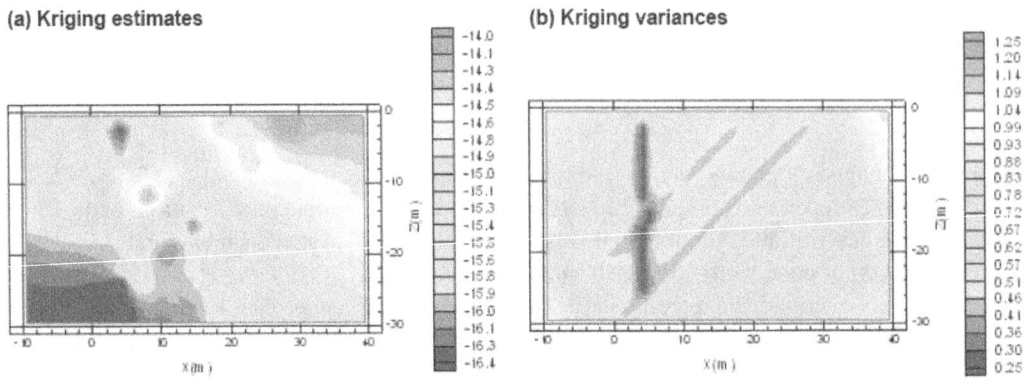

Figure 1. MLBMA kriged estimate and variance of log permeability on a vertical plane at the ALRS.

Table 1. Predictive performance of MLBMA versus that of single candidate models.

	MLBMA	Power model	Exponential model	Exponential model with first-order drift
Predictive log score	31.39	34.11	35.24	33.97
Predictive coverage	87.46	86.49	80.83	83.74

References

Carrera, J., and S.P. Neuman, Estimation of aquifer parameters under transient and steady-state conditions: 1. Maximum likelihood method incorporating prior information, *Water Resour. Res.*, **22**(2), 199_210, 1986a.

Hernandez, A.F., S.P. Neuman, A. Guadagnini, and J. Carrera-Ramirez, Conditioning steady state mean stochastic flow equations on head and hydraulic conductivity measurements, 158-162, *Proc. 4ᵗʰ Intern. Conf. on Calibration and Reliability in Groundwater Modelling (ModelCARE 2002)*, edited by K. Kovar and Z. Hrkal, Charles University, Prague, Czech Republic, 2002.

Hernandez, A.F., S.P. Neuman, A. Guadagnini, and J. Carrera, Conditioning mean steady state flow on hydraulic head and conductivity through geostatistical inversion, *Stochastic Environmental Research and Risk Assessment*, 17(5), 329-338, DOI: 10.1007/s00477-003-0154-4, 2003.

Hoeting, J.A., D. Madigan, A.E. Raftery, and C.T. Volinsky, Bayesian model averaging: A tutorial, *Statist. Sci.*, **14**(4), 382–417, 1999.

Kashyap, R.L., Optimal choice of AR and MA parts in autoregressive moving average models, *IEEE Trans. Pattern Anal. Mach. Intel. PAMI*, **4**(2), 99–104, 1982.

Neuman, S.P., Accounting for conceptual model uncertainty via maximum likelihood model averaging, 529–534, *Proc. 4ᵗʰ Intern. Conf. on Calibration and Reliability in Groundwater Modeling (ModelCARE 2002)*, edited by K. Kovar and Z. Hrkal, Charles University, Prague, Czech Republic, 2002.

Neuman, S.P., Maximum likelihood Bayesian averaging of alternative conceptual-mathematical models, *Stochastic Environmental Research and Risk Assessment*, 17(5), 291-305, DOI: 10.1007/s00477-003-0151-7, 2003.

Neuman, S.P. and P.J. Wierenga, A Comprehensive Strategy of Hydrogeologic Modeling and Uncertainty Analysis for Nuclear Facilities and Sites, NUREG/CR-6805, U.S. Nuclear Regulatory Commission, Washington, DC, 2003.

Samper, F.J. and, S.P. Neuman, Estimation of spatial covariance structures by adjoint state maximum likelihood cross-validation: 1. Theory, *Water Resour. Res.*, **25**(3), 351–362, 1989a.

Development of a Unified Uncertainty Methodology

Philip D. Meyer[1], Ming Ye[1], and Shlomo P. Neuman[2]

[1]Pacific Northwest National Laboratory, philip.meyer@pnl.gov, and ming.ye@pnl.gov
[2]University of Arizona, neuman@hwr.arizona.edu

Multimedia environmental modeling applications generally involve estimating contaminant transport and exposure via complex exposure pathways over a long time period. For example, the primary regulatory criterion for license termination at sites licensed by the U.S. Nuclear Regulatory Commission (NRC) is a maximum dose for the period up to 1000 years from the time of decommissioning. The long regulatory time period and complex transport processes involved in such modeling are often compounded by limited site-specific characterization data. The combination of these factors can result in significant uncertainty in estimates of regulatory quantities such as dose. We are developing a methodology for the comprehensive assessment of hydrogeologic uncertainty in dose modeling. Objectives are that the methodology be applicable to sites with very limited data and to sites with detailed characterization, that it be capable of being applied whether the models used are complex or simplified, and that the methodology should systematically and consistently account for three broad classes of uncertainty: that associated with model parameters, the conceptual basis of the model, and the scenario to which the model is applied.

Quantification of parameter uncertainty for dose assessments must often deal with very limited observations of site characteristics. Generic and indirect data can be and generally are used to infer site properties (Meyer and Gee, 1999). For example, geologic characteristics may be inferred from analysis of outcrops, hydraulic characteristics may be estimated from soil-textural information, and radionuclide adsorption characteristics may be assigned from a database of values measured at other sites under a variety of conditions. Information from the generic and indirect sources can be used to specify prior parameter distributions that can be updated subsequently in a Bayesian approach using site-specific parameter data (Meyer et al., 1997). When observations of state variables (e.g., hydraulic head, radionuclide concentration) are available at a site, the methodology should use formal calibration methods to improve the prior parameter estimates and update the parameter uncertainty (Hill, 1998; Poeter and Hill, 1998; Doherty, 2002, 2003). We rely on the maximum likelihood method (Carrera and Neuman, 1986) because of its general applicability and its effectiveness relative to other methods (Zimmerman et al, 1998). Monte Carlo simulation is used to propagate parameter uncertainty because of its general applicability. Given the potential computational advantage of stochastic moment methods (Dagan and Neuman, 1997) and recent progress in handling conditions that introduce nonstationarities (Zhang, 2001) the methodology should accommodate these methods as well.

Methods for the quantification of conceptual model uncertainty are much less well established than those addressing parameter uncertainty (Mosleh et al., 1994). In the hydrologic field, these methods include an informal comparison of alternatives (James and Oldenburg, 1997; Cole et al., 2001), the likelihood-based weighting of Beven and Freer (2001), the multimodel ensemble approach of Krishnamurti et al. (2000), and the Bayes Factor approach of Gaganis and Smith (2001). We are using the method of Bayesian model averaging (Draper, 1995; Hoeting et al., 1999) to quantify the effect of conceptual model uncertainty. This method combines parameter and conceptual model uncertainty through a weighted average of predictions from a set of alternative models, with the weights being the probabilities that each alternative model is the correct (true) model. Difficulties in implementing this approach include the computational demand of evaluating the integrals involved, specification of the prior model probabilities, and selecting a set of models that is small enough to be computationally feasible yet large enough to represent the breadth of significant possibilities. The latter two issues are related to the interpretation of model probability (Winkler, 1993), which can be resolved by interpreting model probability in relative terms (e.g., Zio and Apostolakis, 1996). In this case, one must recognize that a model with a large probability may still be a poor model, as measured by its predictive ability. Our methodology uses the maximum likelihood implementation of Bayesian model averaging proposed by Neuman (2003). The crucial step of generating alternative conceptual models uses a set of guidelines articulated by Neuman and Wierenga (2003).

A scenario is defined here as a future state or condition assumed for a system that is the result of an event, process or feature, often imposed by humans (e.g., irrigation schemes and ground-water extraction) but may be natural (e.g., glaciation and flooding), which was not assumed in the initial base case definition of the system and diverges significantly from the initial base case. Scenarios are often considered in a long-time context. Quantification of scenario uncertainty can, in principle, be addressed in a manner similar to conceptual model uncertainty (Draper, 1995).

Uncertainties must be defined on a site-specific basis and the importance of individual sources of uncertainty may vary site by site or even with different objectives at the same site. Sensitivity analysis (determination of the factors that are most important to the prediction uncertainty) is an integral element of an uncertainty assessment (Saltelli et al., 2000a; Helton, 1993). Differential, graphical, and sampling-based methods of sensitivity analysis using results from Monte Carlo simulation and optimized parameter estimation are typically applied. We also plan to investigate the importance of global sensitivity measures (Borgonovo et al, 2003; Saltelli et al., 2000b; McKay, 1995), which partition the total prediction variance according to the contribution of each parameter and that due to interactions between parameters.

Acknowledgement and Disclaimer

The research reported here is supported by the U.S. Nuclear Regulatory Commission's Office of Nuclear Regulatory Research under Job Control Number (JCN) Y6465, and is provided for information purposes only and should not be construed as a formal regulatory position.

This abstract was prepared as an account of work sponsored by an agency of the U.S. Government. Neither the U.S. Government nor any agency thereof, nor any employee, makes any warranty, expressed or implied, or assumes any legal liability or responsibility for any third party's use, or the results of such use, of any information, apparatus, product, or process disclosed in this publication, or represents that its use by such third party would not infringe privately owned rights.

References

Beven K.J., and J. Freer (2001). Equifinality, data assimilation, and uncertainty estimation in mechanistic modelling of complex environmental systems using the GLUE methodology, *J. Hydrology*, 249:11–29.

Borgonovo, E., G.E. Apostolakis, S. Tarantola, and A. Saltelli (2003). Comparison of global sensitivity analysis techniques and importance measures in PSA, *Reliability Engineering and System Safety*, 79:175–185.

Carrera J., and S.P. Neuman (1986). Estimation of aquifer parameters under transient and steady state conditions: 1. Maximum likelihood method incorporating prior information, *Water Resour. Res.*, 22(2):199–210.

Cole, C.R., M.P. Bergeron, C.J. Murray, P.D. Thorne, S.K. Wurstner, and P.M. Rogers (2001). *Uncertainty Analysis Framework – Hanford Site-Wide Groundwater Flow and Transport Model*, PNNL-13641, Pacific Northwest National Laboratory, Richland, Washington.

Dagan G., and S.P. Neuman (eds.) (1997). *Subsurface Flow and Transport: A Stochastic Approach*, Cambridge University Press, Cambridge, United Kingdom.

Doherty, J. (2002). *Manual for PEST*, Fifth Edition, Watermark Numerical Computing, Australia.

Doherty, J. (2003). Ground-water model calibration using pilot points and regularization, *Ground Water*, 41(2):170–177.

Draper, D. (1995). Assessment and propagation of model uncertainty, *J. Roy. Statist. Soc. Ser. B*, 57(1):45-97.

Gaganis, P., and L. Smith (2001). A Bayesian approach to the quantification of the effect of model error on the predictions of groundwater models, *Water Resour. Res.* 37(9):2309–2322.

Helton, J.C. (1993). Uncertainty and sensitivity techniques for use in performance assessment for radioactive waste disposal, *Reliability Engineering and System Safety*, 42:327–367.

Hill, M.C. (1998). *Methods and Guidelines for Effective Model Calibration*, U.S. Geological Survey Water-Resources Investigations Report 98-4005, U.S. Geological Survey, Denver, Colorado.

Hoeting, J.A., D. Madigan, A.E. Raftery, and C.T. Volinsky (1999). Bayesian model averaging: A tutorial, *Statist. Sci.*, 14(4):382–417.

James A.L., and C.M. Oldenburg (1997). Linear and Monte Carlo uncertainty analysis for subsurface contaminant transport simulation, *Water Resour. Res.*, 33(11):2495–2508.

Krishnamurti, T.N., C.M. Kishtawal, Z. Zhang, T. LaRow, D. Bachiochi. E. Williford, S. Gadgil, and S. Surendran (2000). Multimodel ensemble forecasts for weather and seasonal climate, *J. Climate*, 13(23):4196–4216.

McKay, M.D. (1995). *Evaluating Prediction Uncertainty*, NUREG/CR-6311, U.S. Nuclear Regulatory Commission, Washington, D.C.

Meyer, P.D., M.L. Rockhold, and G.W. Gee (1997). *Uncertainty Analyses of Infiltration and Subsurface Flow and Transport for SDMP Sites*, NUREG/CR-6565, U.S. Nuclear Regulatory Commission, Washington, DC (http://nrc-hydro-uncert.pnl.gov/).

Meyer, P.D. and G.W. Gee (1999). *Information on Hydrologic Conceptual Models, Parameters, Uncertainty Analysis, and Data Sources for Dose Assessments at Decommissioning Sites*, NUREG/CR-6656, U.S. Nuclear Regulatory Commission, Washington, DC (http://nrc-hydro-uncert.pnl.gov/)

Mosleh, A., N. Siu, C. Smidts, and C. Lui (eds.) (1994). *Model Uncertainty: Its Characterization and Quantification, Proceedings of Workshop I in Advanced Topics in Risk and Reliability Analysis*, NUREG/CP-0138, U.S. Nuclear Regulatory Commission, Washington, DC.

Neuman, S.P. (2003). Maximum likelihood Bayesian averaging of alternative conceptual-mathematical models, *Stochastic Environmental Research and Risk Assessment* (in press).

Neuman, S.P., and P.J. Wierenga (2003). *A Comprehensive Strategy of Hydrogeologic Modeling and Uncertainty Analysis for Nuclear Facilities and Sites*, NUREG/CR-6805, U.S. Nuclear Regulatory Commission, Washington, DC.

Poeter, E.P., and M.C. Hill (1998). *Documentation of UCODE, A Computer Code for Universal Inverse Modeling*, U.S. Geological Survey Water-Resources Investigations Report 98-4080, 116 pp., U.S. Geological Survey, Denver, Colorado.

Saltelli, A., K. Chan, and E.M. Scott (eds.) (2000a). *Sensitivity Analysis*, John Wiley & Sons LTD, Chichester, England, 475 pp.

Saltelli, A., S. Tarantola, and F. Campolongo (2000b). Sensitivity analysis as an ingredient of modeling, *Statistical Science*, Vol. 15, 4:377–395.

Winkler, R.L. (1993). "Model uncertainty: probabilities for models?," in Mosleh, A., N. Siu, C. Smidts, and C. Lui (eds.), *Model Uncertainty: Its Characterization and Quantification, Proceedings of Workshop I in Advanced Topics in Risk and Reliability Analysis*, NUREG/CP-0138, U.S. Nuclear Regulatory Commission, Washington, DC.

Zhang, D. (2001). *Stochastic Methods for Flow in Porous Media*, Academic Press.

Zimmerman, D.A., G. de Marsily, C.A. Gotway, M.G. Marietta, C.L. Axness, R.L. Beauheim, R.L. Bras, J. Carrera, G. Dagan, P.B. Davies, D.P. Gallegos, A. Galli, J. Gómez-Hernández, P. Grindrod, A. L. Gutjahr, P.K. Kitanidis, A.M. Lavenue, D. McLaughlin, S.P. Neuman, B.S. RamaRao, C. Ravenne, and Y. Rubin (1998). A comparison of seven geostatistically based inverse approaches to estimate transmissivities for modeling advective transport by groundwater flow, *Water Resour. Res.*, 34(6):1373–1413.

Zio, E. and G.E. Apostolakis (1996). Two methods for the structured assessment of model uncertainty by experts in performance assessments of radioactive waste repositories, *Reliability Engineering and System Safety*, 54:225–241.

6

SESSION 4:

PARAMETER ESTIMATION, SENSITIVITY AND UNCERTAINTY APPROACHES — *APPLICATIONS AND LESSONS LEARNED*

6.1 Overview and Summary

Editors: Bruce Hicks and George Leavesley

Presentations addressed the issue of practical multimedia environmental modeling, with emphasis on the need for both case studies to reveal flaws in the understanding of critical processes, and advanced computational capabilities to permit the complex models to be run. The issue of model complexity generated lengthy discussion, with several speakers strongly endorsing the principle of parsimony – in essence, do not make a model more complicated unless there is sound reason to do so. A complicated model does not necessarily give better answers than a simpler model. (In practice, a strong reason for much model complexity is often to avoid the criticism of specialists.) It is comparison against data that will show whether increasing model complexity results in improved model performance. Without relevant data, the benefits of increased complexity cannot be demonstrated.

There is need to consider both deterministic and probabilistic approaches. In some circumstances, the former will work better than the latter. In other situations, the opposite may be true. As yet, there is little confidence in the ability to determine the point at which probabilistic (and/or empirical) methods will start to work better than deterministic, but in general, it is accepted that the capability that is being sought must have aspects of both approaches. Deterministic approaches often incorporate a "margin of error," which is a step toward linking with probabilistic methods. Regulatory systems do not yet accept probabilistic guidance with confidence. To assist in communicating the results of probabilistic analyses, improved methods are needed for depicting and characterizing uncertainty. In the absence of a widely accepted communication protocol, extensive and continuing dialogue is usually necessary.

It was pointed out that modern modeling methods have made largely obsolete the historic standards and criteria used in decision-making. Model capabilities have grown. Regulatory systems tend to change far more slowly. Reliance on Total Maximum Daily Loads (TMDLs) has not alleviated the concerns. Lengthy discussion on TMDLs revealed that several difficulties remain to be addressed. For example, lags between pollution inputs and consequent effects need to be taken into account. TMDLs should be adaptive. In general, a monitoring program is needed to support them.

To address multimedia questions, a large number of process-related models is usually required, and these need to be linked in a coherent fashion. Sometimes, it is appropriate to use simplified descriptions, so as to impose some balance in the way that key processes are addressed while retaining the detail necessary to accomplish specified goals. Even in the case of such "engineering models," there must be some description of all of the many contributing processes.

No matter what modeling approach is adopted, it is important to consider whether the predictions can be confirmed with system behavior measurements. Under the best of circumstances, our goal would be to only predict things we can explicitly measure, but the decision-based reality we face today imposes a need to use models to predict a variety of events and consequences that may not be realistically measurable in time (i.e., from either feasibility or prohibitive cost perspectives). Therefore, we must necessarily use great caution in moving down the road of model prediction with minimal to no confirmatory measurements. In practice, all opportunities to evaluate the performance of models should be taken, in circumstances that parallel those of their intended application. Moreover, it is misleading to construct a final answer by imposing a conservative assumption for each of these sequential process sub-models. There are formal methods for propagating uncertainties through complex modeling systems. These need to be utilized. Moreover, it must be recognized that natural variability plays an important part and must be taken into account. Once a probabilistic methodology is adopted, evaluation of the products presents difficulties not common in the case of fully deterministic approaches.

In all cases, documentation of models and steps taken to refine them is critical, especially in regard to the way in which uncertainties are addressed and propagated. Too often, written documentation lags far behind.

There is a propagation of errors that parallels the propagation of uncertainties through the modeling systems used to address multimedia concerns. Detection of such errors requires close attention to assumptions, descriptions of processes, and coding. This is one component of a model evaluation procedure that is critical to any effort to gain acceptance for the products that are developed. The most visible step in this procedure is clearly a test against observations, but clearly code examination and formal peer review are also critical. It was pointed out that we learn most from models that disagree with observations.

Multimedia models of contemporary times consider different media, a variety of pathways, and many different receptors. The extension to multiple stressors has yet to take place, yet it is clearly evident that today's environment is increasingly at risk, not from one but from a large number of threats, any one of which may prove deadly in some specific set of circumstances.

Data with which to evaluate model performance are exceedingly rare. Watershed data are especially desired, collected so that models can be tested diagnostically. Testing on the basis of agreement with observations is sometimes risky, because it avoids the intermediate steps to provide assurance that the correct answers were obtained correctly. In reality, there are few relevant measurements being made routinely. There is special need for diagnostic data to be obtained in intensive studies. To this end, a partnership with the experimental community is sought. One question to be addressed in such a partnership relates to the difficulty in quantifying uncertainties in the absence of data.

There is an international effort to construct a Global Observing System. The multimedia modeling community needs to have input into the international monitoring design process.

Watersheds present excellent opportunities for evaluating multimedia models. Flood data are often especially useful.

Probabilistic Risk Assessment for Total Maximum Daily Surface-Water Loads (TMDLs)

Kenneth H. Reckhow, Mark E. Borsuk, and Craig A. Stow

Nicholas School of the Environment and Earth Sciences
Duke University, Durham, North Carolina USA 27708-0328

Reckhow@duke.edu

TMDL assessment and forecasting may require characterization of a number of physical, chemical, and biological factors linking pollutant sources to water quality criteria. For example, the symptoms of coastal and estuarine eutrophication are the result of several interacting processes operating at multiple spatial and temporal scales. Thus, submodels developed to appropriately represent each of these processes may not easily be combined into a single predictive model that supports quantification of prediction uncertainties for risk assessment. We suggest that Bayesian networks provide a possible solution to this problem. The graphical structure of the Bayes net explicitly represents cause-and-effect assumptions between system variables, expressed in a probabilistic manner. These assumptions allow the complex causal chain linking management actions to ecological consequences to be factored into an articulated sequence of conditional probabilities. Each of these relationships can then be quantified independently using an approach suitable for the type and scale of information available. Probabilistic functions describing the relationships allow key known or expected mechanisms to be represented without the full complexity, or information needs, of highly detailed reductionist models.

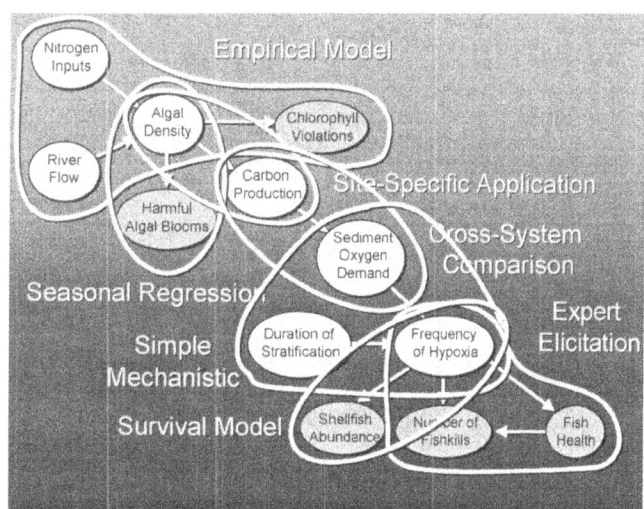

Figure 1: The Neuse River Estuary Bayes Net Model

To demonstrate the application of the approach, we develop a Bayesian network representing eutrophication in the Neuse River Estuary of North Carolina from a collection of previously published analyses. Relationships among variables were quantified using a variety of methods, including process-based models statistically fit to long-term monitoring data, Bayesian hierarchical modeling of cross-system data, multivariate regression modeling of mesocosm experiments, and probability judgments elicited from scientific experts (Figure 1). We use the fully quantified model to generate probabilistic predictions of ecosystem response to alternative nutrient management strategies in the development of a TMDL for nitrogen in the Neuse River Estuary (Figure 2). The probabilistic nature of the Bayes net model provided the basis for the margin of safety estimation (Figure 3); further, it served to enlighten stakeholders concerning the limitations of water quality forecasting.

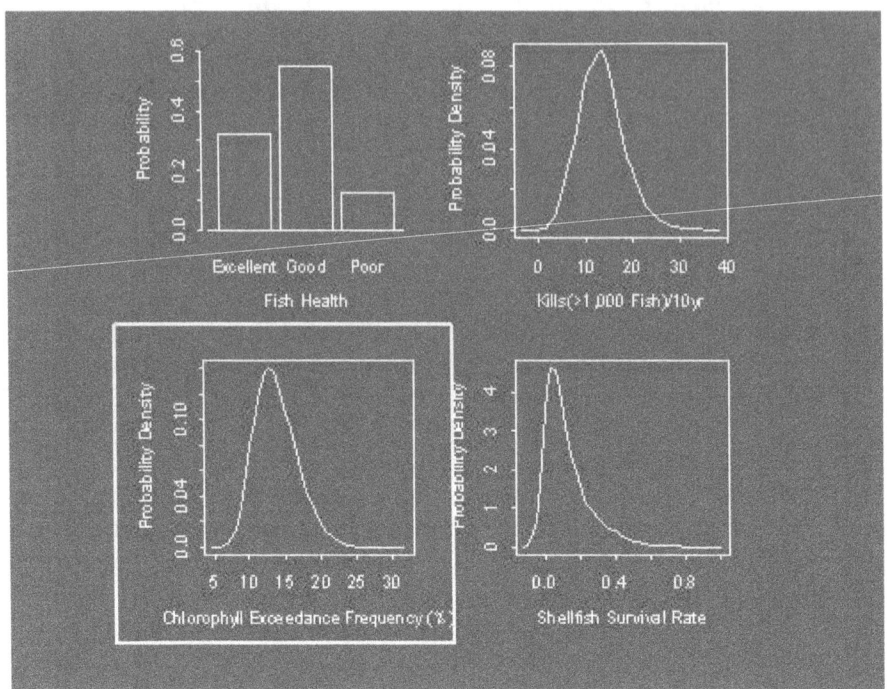

Figure 2: Probabilistic Predictions from Neuse River Estuary Bayes Net Model

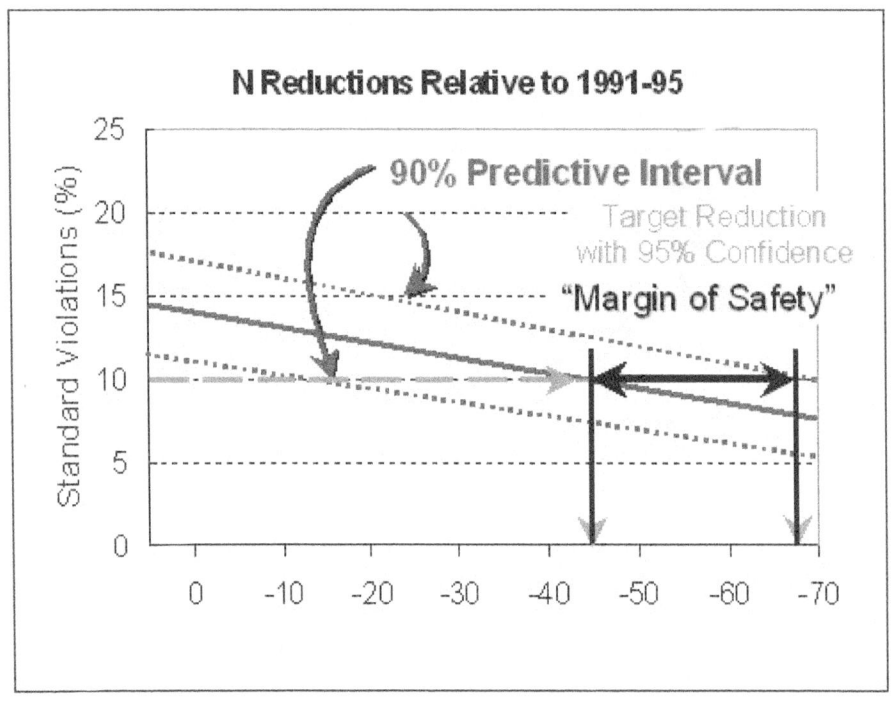

Figure 3: Margin of Safety Estimation

References

Borsuk, M.E., C.A. Stow, and K.H. Reckhow. 2003. An integrated approach to TMDL development for the Neuse River Estuary using a Bayesian probability network model (Neu-BERN), *Journal Water Resources Planning and Management. 129*:271–282.

Borsuk, M.E., C.A. Stow, and K.H. Reckhow. 2002. Predicting the frequency of water quality standard violations: A probabilistic approach for TMDL development. *Environmental Science and Technology. 36*:2109–2115.

Borsuk, M.E., D. Higdon, C.A. Stow, and K.H. Reckhow. 2001. A Bayesian Hierarchical Model to Predict Benthic Oxygen Demand from Organic Matter Loading in Estuaries and Coastal Zones. *Ecological Modelling. 143*:165-181.

Borsuk, M., R. Clemen, L. Maguire, and K. Reckhow. 2001. A Multiple-Criteria Bayes Net Model of the Neuse River Estuary. *Group Decision and Negotiation. 10*:355–373.

Reckhow, K.H. 2003. On the need for uncertainty assessment for TMDL modeling and implementation. *Journal Water Resources Planning and Management. 129*:245–246.

Stow, Craig A., Chris Roessler, Mark E. Borsuk, James D. Bowen, and Kenneth H. Reckhow. 2003. A Comparison of Estuarine Water Quality Models for TMDL Development in the Neuse River Estuary. *Journal Water Resources Planning and Management. 129*:307–314.

Stow, C.A., M.E. Borsuk, and K.H. Reckhow. 2002. Nitrogen TMDL development in the Neuse River Watershed: an imperative for adaptive management. *Water Resources Update.* The Universities Council on Water Resources. *122*:16–26.

A Stochastic Risk Model for the Hanford Nuclear Site

Paul W. Eslinger

Pacific Northwest National Laboratory, paul.w.eslinger@pnl.gov

The U.S. Department of Energy (DOE) faces many decisions regarding future remedial actions and waste disposal at the Hanford Site in southeast Washington State. A new software framework, the System Assessment Capability (SAC), has been developed to provide the DOE with the means to predict cumulative impacts of waste disposal and remediation plans accounting for hundreds of individual disposal locations on the 1517-square-kilometer Hanford Site. To support decision-making in the face of uncertainty, the SAC was built as a stochastic framework so that uncertainty in predictions could be based on uncertainty in input parameters and conceptual models. The code is implemented in the FORTRAN 95 language and is designed to run on a 132-CPU Linux® cluster.

The SAC simulates contaminant release, migration, and fate from the initiation of Hanford Site operations in 1944 forward. It illustrates historical and near-term influences on long-term risk and impact and, therefore, provides an opportunity to history match to observed events. The design separates the environmental and risk/impact simulations, and archives the environmental results so that the DOE, regulatory agencies, Tribal Nations, and stakeholders may explore multiple risk/impact scenarios. Impacts are estimated for four components of the environment and society: ecological health, human health, economic conditions, and cultural resources. The SAC is able to model multiple contaminants at 1,000 or more waste or disposal sites for a period of 10,000 years or longer. It has been designed to simulate a deterministic case as a single stochastic realization.

An initial run of the SAC using 10 contaminants has been completed and documented (Bryce *et. al*, 2002). The human impacts analysis examined exposure scenarios ranging from the ingestion of contaminated water to farming or recreational activities on the Hanford site and in the Columbia River. The economics impacts model examined potential deviations from the current regional economy due to future migration of contaminants. The cultural model examined the impacts of the contaminated groundwater on the newly created Hanford national monument. The ecological impacts estimation uses a food-web approach that analyzed the effects on 57 representative species along the Columbia River from Vernita Bridge to McNary Dam. The highest impacts are estimated to occur near the site of retired reactors. In general, the groundwater plumes developed in the model are similar to the historical record of groundwater contamination and contamination in the Columbia River. The uncertainty analysis shows a magnitude spread of about 2 orders of magnitude in most estimated impact metrics.

The initial run was a proof-of-principle demonstration of the modeling approach. A revised version of the code is in the final testing phase and will provide a tool suitable for regulatory applications requiring or benefiting from a site-wide assessment of risks and impacts associated with contaminants remaining at the Hanford Site after closure. An assessment recently completed is documented in the cumulative impacts section of the Hanford Solid Waste Environmental Impact Assessment. Another assessment in progress is an update to the Composite Analysis (DOE Order 435.1) for the Hanford Site.

Data collection for any large-scale environmental simulation is a time-consuming process. We have encountered three general data difficulties while conducting stochastic simulations. First, radioactive waste data collected for other purposes often suffers from ultra-conservative approaches or interpretations. The desire for a safety margin in one model can lead to unrealistically high impact estimates when coupled with another model. Second, it is difficult or expensive to incorporate alternative conceptual models in a stochastic simulation. Alternative conceptual models are important in some cases. For example, contaminants from the 200 East Area on the Hanford site may move in different directions in the groundwater depending on the conceptual model. Finally, developing realistic statistical distributions for input data is difficult. One area of interest is the parameters in the van Genuchten and Maulem models used in the unsaturated zone hydrologic model. A specialized sampling scheme has been developed to reject combinations of input parameters that lead to unrealistic outcomes.

References

Bryce, R.W., C.T. Kincaid, P.W. Eslinger, and L.F. Morasch (eds.). 2002. An Initial Assessment of Hanford Impact Performed with the System Assessment Capability. PNNL-14027, Pacific Northwest National Laboratory, Richland, Washington.

Eslinger. P.W., D.W. Engel, L.H. Gerhardstein, C.A. Lo Presti, W.E. Nichols, D.L. Strenge. June 2002. User Instructions for the Systems Assessment Capability, Rev. 0, Computer Codes, Volume 1: Inventory, Release, and Transport Modules. PNNL-13932-Volume 1, Pacific Northwest National Laboratory, Richland, Washington.

Eslinger, P.W., C. Arimescu, B.A. Kanyid, and T.B. Miley. June 2002. User Instructions for the System Assessment Capability, Rev. 0, Computer Codes. Volume 2: Impact Modules. PNNL-13932-Volume 2, Pacific Northwest National Laboratory, Richland, Washington.

Kincaid, C.T., P.W. Eslinger, W.E. Nichols, A.L. Bunn, R.W. Bryce, T.B. Miley, M.C. Richmond, S.F. Snyder, and R.L. Aaberg. 2000. System Assessment Capability, Rev. 0, Assessment Description, Requirements, Software Design, and Test Plan. BHI-01365, Draft A, Bechtel Hanford, Inc., Richland, Washington.

Acknowledgments

This project was a large team effort. The major team participants are listed as coauthors of one or more of the four references provided above. This work was performed by Battelle for the U.S. Department of Energy under contract DE-AC06076RL01830, in partnership with Fluor Hanford and Bechtel Hanford, Inc.

National-Scale Multimedia Risk Assessment for Hazardous Waste Disposal

Justin E. Babendreier

Ecosystems Research Division, National Exposure Research Laboratory,
Office of Research and Development, U.S. Environmental Protection Agency
Athens, Georgia   30605

babendreier.justin@epa.gov

While there is a high potential for exposure of humans and ecosystems to chemicals released from a single hazardous waste site, the degree to which this potential is realized is often uncertain. Conceptually divided among parameter, model, and modeler uncertainties imparted during simulation, inaccuracy in model predictions result principally from lack of knowledge and data. In comparison, sensitivity analysis can lead to a better understanding of how models respond to variation in their inputs, which in turn can be used to better focus laboratory and field-based data collection efforts on processes and parameters that contribute most to uncertainty in outputs. We generally seek to both describe uncertainty for the current state of science and data, and, further, to ascertain a prioritized agenda for its reduction. The former allows for the critical task of making informed wastestream management decisions in the present, and the latter, ideally, drives the research planning process. For environmental regulation, these two elements, action and continued research investigation, represent encompassing statements describing the daily execution of EPA's primary mission to protect human health and the environment. It is a combined process deeply rooted in the fundamental engineering principle of cost-benefit analysis.

Multiplicity (An Operative Concept for Future Risk Assessment Paradigms)

As we rapidly push forward to integrate multimedia, multipathway, multireceptor, multi-contaminant, and multi-scale risk assessments associated with hazardous waste disposal, we are invariably led to an increasingly complex problem statement and modeling paradigm. Complexity of the problem statement increases substantially in concurrently addressing risks to both human and ecological populations and their associated subpopulations (e.g., "high end" sensitive receptors, etc.). Further compounding national management approaches for various hazardous wastestreams, national assessment strategies, derived from multiple, site-based risk assessments, present even greater challenges in evaluating confidence in model-based forecasts of population protection. Due to its inherent abstraction, national management strategies also present increasing difficulty in communicating risk to both decision-makers and stakeholders, while overlooking alternative efficiencies possibly available, though at far greater management cost, in dealing with risk on more resolved spatial scales. Depending on the waste constituent of interest, protection forecasts will typically also need to span years to thousands of years.

The FRAMES-3MRA Modeling System (Marin et al., 2003; Babendreier, 2003)

Residing within the Framework for Risk Analysis in Multimedia Environmental Systems (FRAMES), the Multimedia, Multipathway, and Multireceptor Risk Assessment (3MRA) modeling system was developed by EPA for use in assessing risks from hazardous waste disposal. The 3MRA modeling system, basically a screening level risk assessment technology, includes a set of 17 science

137

modules that collectively simulate release, fate and transport, exposure, and risk associated with hazardous contaminants disposed of in land-based waste management units (WMUs). The 3MRA model currently encompasses 966 input variables, over 185 of which are explicitly stochastic. 3MRA starts with a wastestream concentration in one of five WMU types, estimates the release and transport of the waste constituent chemical or metal throughout the environment, and predicts associated exposure and risk. 3MRA simulates multimedia (air, water, soil, sediments) fate and transport, multipathway exposure routes (food ingestion, water ingestion, soil ingestion, air inhalation, etc.), multireceptor exposures (resident, gardener, farmer, fisher, ecological habitats and populations, all with various cohort considerations), and resulting risk (human cancer and non-cancer effects, and ecological population and community effects). At the heart of the assessment approach, is the organization of available data sets into national, regional, and site-based databases, and meteorological and chemical property databases. Incorporating landfills, waste piles, aerated tanks, surface impoundments, and land application units, the current site-based data is comprised of 201 statistically sampled national facilities representing 419 site-WMU combinations, and a chemical property database representing 43 organic chemicals and metals.

National Risk Assessment Problem Statement Formulation for Hazardous Waste Disposal

A key question 3MRA is capable of answering may be stated as follows: At what wastestream concentration (C_w) will wastes, when placed in a non-hazardous WMU over the unit's life, result in:

- Greater than A% of the people living within B distance of the WMU with a risk/hazard of C or less, and

- Greater than D% of the habitats within E distance of the WMU with an ecological hazard of F or less,

- At G% of facilities nationwide?

A probability (H) may also be assigned to empirical input uncertainty associated with the derived protection profile for percentiles of the target population or subpopulations (e.g., uncertainty in C_w). Furthermore, a probability (I) may be assigned to the simulation-derived empirical output uncertainty associated with the derived protection profile for percentiles of the target population or subpopulations. Defining the assessment profile (A, B, C, D, E, F, G, H, I), 3MRA embodies an integrated, probabilistic risk assessment strategy for protection of both ecological and human health. The above construct (A, B, C, D, E, F, G) imparts a statement of variability in the output, (H) imparts uncertainty due to lack of knowledge and data (i.e., empirical input uncertainty), and (I) imparts empirical uncertainty due purely to computational constraints in simulating output distributions [e.g., Monte Carlo Simulation (MCS) error].

Qualitative and Quantitative Model Evaluation Approaches

Assessment of the effects of empirical uncertainty and variability in model inputs upon output, derived from their explicit representations in model inputs, generally first involves the propagation of both through the model. It is also often desired to apportion variance in inputs to variance in outputs.

Aspects of sensitivity for a given model may be evaluated through a wide array of computational techniques, for example, screening methods, local differential-based methods, and global methods (Saltelli et al., 2000). In addition to the variance-based global sensitivity methods outlined in

138

Saltelli et al. (2000), which provide the ability to quantitatively relate variance in input to variance in output, there are equally provocative schemes (Funtowicz and Ravetz, 1990) to be investigated that more fully characterize elements of uncertainty, reaching well beyond the quantifiable, commonly applied (multi-dimensional) Monte Carlo-based probabilistic assessments (Cullen and Frey, 1999). In the NUSAP (Numeral, Unit, Spread, Assessment, and Pedigree) scheme of Funtowicz and Ravetz (1990), for example, uncertainty is constructed along a continuum of familiar, quantitative information, as well as less familiar, qualitative information that asserts a level of confidence in the former. Together, the NUSAP entities (van der Sluijs, 2003) impart a deep structure of quality assurance in the information system otherwise historically represented by a model's prediction and the best of intentions.

Though done outside the direct guidance of the NUSAP method as a model evaluation and quality assurance guidance tool, in retrospect, the 3MRA development, documentation, and peer-review process undertook these major steps along similar lines. To sustain our current course of evaluating ever more complex questions through use of increasingly complex models, variability, uncertainty, and sensitivity analyses will likely continue to rely on application of sampling-based techniques (e.g., MCS). The future will also continue to see advances in methodological approach, and modelers will predictably desire to apply these computationally demanding procedures in a timely fashion (Beck, 1999).

The Model Validation Paradox

Extending beyond a simplistic, unworkable view of retrospectively oriented model performance validation exercises rooted in history matching, components of model evaluation for 3MRA are viewed here as inextricably linked to a familiar concept of quality assurance in product (tool or technology) design (Beck et al., 1997). "Use" in regulatory decision-making typically implies the final exercise of the model as a forecast of some subjectively determined protection level of human health and the environment. Only direct auditing of future attainment of the desired risk assessment objective (e.g., a certain level of protection achieved by a specific waste constituent management strategy over time) could begin to approach full illumination of the model's success, and our grasp of science involved. Even then, such a determination, if it were feasible to construct, would realistically remain, after the fact, a substantially subjective conclusion for complex problem statements such as those addressed by 3MRA.

For example, it is arguably untenable that one could go about verifying 30 years from now that 30 years of past waste management practices at 10,000 waste management facilities across the United States have imparted a specific increased risk of cancer for 300 million human beings—or even 100,000 for that matter. That there is inherent subjectivity in any post-audit determination becomes increasingly apparent as we add to this the perspective of auditing some chosen level of protection for ecological systems from the same waste management practices.

Our focus for the time being is on a more attainable, tactical challenge of evaluating the 3MRA technology for a specific use in the present. The present use is the task of predicting future system behavior under novel conditions—an unobservable future for the time being. The problem of reaching a satisfactory, empirically based measure of validation in the present is restrained by two dilemmas: (1) the future truth we seek is paradoxically unobservable in the present, and (2) subjective decision variables used in complex problems, such as exposure and risk assessments, are realistically unobservable in the present and future. Fundamentally, in regulatory endeavors, one will face an unavoidable dilemma of extrapolation toward unobservable futures. As a performance validation measure of 3MRA, we build upon the works of Young, Hornberger, Spear, Beck, Chen, and Osidele (Osidele and Beck, 2003) in developing the notion of a model having maximum relevance to the performance of a specific task, through use of Regional Sensitivity Analysis (RSA) and Tree-Structured Density Estimation (TSDE), broadening the discussion of model validation into one of quality assurance in environmental forecasting.

Quality Assurance in Environmental Forecasting

In formulating regulation, the agency is increasingly held accountable today to formally demonstrate that the underlying science and data used are, to the extent practical, accurate, reliable, unbiased, and reproducible (U.S. EPA, 2002). Further, regulators must establish that the presentation of information available is sufficiently comprehensive, informative, and understandable so as to allow the public to understand the risk assessment methodology and populations being considered, and the agency's plans for identifying and evaluating the uncertainty in risks. In summarizing the national problem statement for risk assessment of hazardous waste disposal, we should first acknowledge that evaluating uncertainty and sensitivity in environmental models can be a difficult task, even for low-order, single-medium constructs driven by a unique set of site-specific data.

Quantitative assessment of integrated, multimedia models that simulate hundreds of sites, spanning multiple geographical and ecological regions will ultimately require a comparative approach using several techniques, coupled with sufficient computational power. The challenge of examining ever more complex, integrated, higher-order models is formidable in regulatory settings applied on national scales that must ensure protection of humans and ecology, while preserving the economic viability of industry. We are, thus, increasingly driven to provide enhanced confidence and a technical basis for regulatory decisions through integrated, "full-service" modeling, essentially bringing science and its uncertainties directly into regulation. In actual fact, a statement of the quality assurance in a model's use for its intended purpose is no longer optional, but indeed requisite.

Achieving adequate quality assurance in modeling, in essence, requires a battery of tests designed to establish the model's validity, trustworthiness, and relevance in performing a prospective task of prediction (Chen and Beck, 1999). Together with peer review and iterative application, this process derives qualitative and quantitative information on various aspects of simulation science and model verification, validation, assessment (and separation) of variability and uncertainty in inputs, assessment of model structure errors, and the identification of the sensitivity of model output to key model inputs.

On the subject of determining sufficient performance validation for novel conditions, the crux of the matter lays in developing a fully consistent problem statement, the reality of reaching a successful description of model validation for a given purpose will require not only a statement of the desired risk assessment objective, but also a description of undesirable outcomes of performance (Beck et al., 1997; Burns et al., 1990; Burns, 1983, 2001). Thus, minimum external model validation is gauged by its intended use and, on some level, can be formulated as a tolerance for failure.

Model Evaluation Strategy for 3MRA

In addition to compositional validation (Beck et al., 1997) (e.g., verification), which has included extensive peer reviews of science-module constituent hypotheses and their integration, and extensive module and system-level testing, the 3MRA model evaluation plan also comprises three additional, major tasks:

Performance uncertainty analysis (UA$_p$) basically entails propagation of input uncertainty and variability through the modeling system, while also addressing output sampling error (OSE) associated with computational limitations of the sampling-based MCS strategy. It is performed using a pseudo second-order analysis to address empirical input uncertainty and OSE. Depending on outcomes of the sensitivity analyses (SA), a limited, yet more broadly scoped second-order analysis could possibly be undertaken, to the degree feasible. Such an analysis might, for example, further address uncertainty in the empirical distribution specifications associated with sample measurement error (SME) and input sampling error (ISE) for extremely sensitive (i.e., key) model inputs, provided suitable information could be made available to form the analysis. For the complexity represented by 3MRA, absolute model error (ME) cannot be formally quantified at this time due to an overall lack of knowledge and data available that would make such an effort meaningful.

The formal analysis of 3MRA predictive uncertainty focuses on empirical uncertainty derived from (1) the use of *variable and certain* national, regional, and site-based random input variables describing national and regional variability of various model inputs, where uncertainty is imparted in their use to describe individual site-based assessments that make up the national assessment strategy; and (2) the use of *constant and uncertain* national, regional, and site-based random input variables, for example, that characterize wastestream properties or various chemical properties. Approaches allow for separation of empirical-based uncertainties from natural variability derived from inputs measured at various sites, as represented in the regional and site databases. These are, of course, all tentative designations that could be further expanded upon with additional data collection and/or model input characterization.

System-level sensitivity analysis (SA) basically explores the mapping $[\mathbf{x}_k, \mathbf{y}(\mathbf{x}_k)]$ through use of several analytical techniques, identifying key, important, and redundant model inputs. SA to be conducted for this purpose will enhance both compositional and performance validation aspects for the modeling system. The latter (i.e., an aspect of performance uncertainty analysis) is reflected upon as a qualification of the importance of accurately quantifying input uncertainty in support of the final UA_p. The former (i.e., compositional validation) represents additional activity supporting module-level and system-level modeling system verification (through identification of unexpected model output behavior over the allowable ranges of inputs; e.g., programming errors, discontinuities, non-linearity, non-monotonicity etc.). For SA work, familiar regression/correlation-based procedures (Helton and Davis, 2000; Kleijnen and Helton, 1999) will be employed, in addition to use of the RSA and TSDE global-based sensitivity analysis techniques.

Sensitivity-analysis-based performance validation (SA$_p$) involves an assessment of a "prior" validity through the execution of a univariate RSA procedure and as feasible, through the use of the multivariate TSDE procedure, both to be realized as an assessment of the model's maximum relevancy in predicting model behavior for various population percentiles (Beck et al., 1997; Chen and Beck, 1999; Beck and Chen, 2000).

Interpretation of 3MRA Site-Based National Realizations

3MRA output is essentially based on one or more deterministic runs of the modeling system. For the national assessment, a site-based analysis of 201 sites is formed from queries from the national, regional, and site-based 3MRA databases, site-by-site, to form the necessary modeling system inputs. The national assessment is constructed from repeated collections of potential outcomes across these 201 representative sites. In interpreting risk analysis results of the 3MRA national study, a cardinal rule of risk analysis modeling subscribed to here is summarized by Vose (2000), inferring that every 3MRA national realization represents a national scenario that could physically occur. This distinction is quite important to the interpretation held for output data generated by 3MRA for the national study. In summarizing this strategic point, we view that a single, national realization of the representative 201 sites represents a potential outcome (or sample) of future waste management conditions, nationally, with some probability (i.e., uncertainty) of occurrence.

The aspect of national, site-based assessments, such as that discussed here for 3MRA, imposes unique, practical challenges in assignment of model inputs to various cases of total uncertainty and subsequent interpretation of modeling system output. This is because of the complexity normally imposed by site-specific studies, commingled with (1) the aspect of rolling up risk analyses across multiple risk assessments of single sites, all deriving data, in sometimes different fashion, from various scaled databases (i.e., site-based, regional, and national); and (2) the onus of evaluating how "variability" of the true national target population is actually expressed within the site-based sampling design, model simulation design and, ultimately, the problem statement. A fundamental aspect of interpretation of 3MRA model output is borne out of the idea that, in context of the model design and database construction, the true target population represents a collection of an infinite (or at least an extremely large) number of sites that would be needed to embody the entire potential of national and regional variability. In reality, the decision-maker is faced with the perspective that over any time frame, only portions of this potential variability will actually be realized. It is this limited potential, as a statement of probability (i.e., uncertainty), that decisions of population protection should actually be based upon.

A Novel Hardware and Software Computational Strategy for Windows-Based Models

A characteristic of uncertainty analysis (UA) and sensitivity analysis (SA) for very-high-order models (VHOMs) like 3MRA is their need for significant computational capacity to perform relatively redundant simulations. We refer to this UA/SA problem statement as an embarrassingly parallel computational problem, in juxtaposition to massively parallel computational techniques (Brightwell et al., 2000). While UA/SA is emerging as a critical area for environmental model evaluation, proper evaluation of Windows-based models have been limited by a lack of supercomputing capacity. Equally, higher-order UA/SA algorithms warrant investigation to determine their efficacy in establishing requisite confidence in the use of VHOMs for regulatory decision-making.

Design of **SuperMUSE** (Babendreier and Castleton, 2002; Babendreier, 2003), a 215 GHz PC-based, Windows-based **Super**computer for **M**odel **U**ncertainty and **S**ensitivity **E**valuation, is described. 3MRA model results are presented here for an uncertainty analysis example of benzene disposal using 3MRA that shows the relative importance of various exposure pathways in driving risk levels for ecological receptors and human health, exemplifying aspects of the national-scale assessment methodology. As an example of compositional validation work completed, using SuperMUSE, over 40 million individual 3MRA model simulations have been conducted to date, where average model run times are on the order of 2 minutes. Convergence in output sampling is expected to require on the order of millions to tens of millions of model runs for seven chemicals currently under study. Generally, overhead in parallel processing is negligible and the approach is fully scalable.

References

Babendreier J.E., and K.J. Castleton. (2002). Investigating Uncertainty and Sensitivity in Integrated, Multimedia Environmental Models: Tools for FRAMES-3MRA. In *Proc. of 1st Biennial Meeting of International Environmental Modeling and Software Society*, (2) 90–95, Lugano, Switzerland.

Babendreier, J.B. (2003). The Multimedia, Multipathway, Multireceptor Risk Assessment Modeling System (FRAMES-3MRA Version 1.0) Documentation. Volume IV: Evaluating Uncertainty and Sensitivity. Draft SAB Review Report: EPA530/D/03/001d, U.S. Environmental Protection Agency Office of Solid Waste and Office of Research and Dev., Washington DC, http://www.epa.gov/ceampubl/mmedia/3mra/index htm. See also Volumes I, II, III, and V: EPA530/D/03/001a:b:c:e.

Beck, M.B., and J. Chen. (2000). Assuring the Quality of Models Designed for Predictive Tasks. In *Sensitivity Analysis* (A. Saltelli, K. Chan, and E.M. Scott, eds.), John Wiley & Sons: West Sussex, England, pp. 401–420.

Beck, M.B., J.R. Ravetz, L.A. Mulkey, T.O. Barnwell. (1997). On the Problem of Model Validation for Predictive Exposure Assessments. *Stochastic Hydrology and Hydraulics*, 11:229–254.

Beck, M.B. (1999). Coping With Ever Larger Problems, Models, and Databases. *Wat. Sci. Tech.*, 39 (4):1–11.

Brightwell, R., L.A. Fisk, D.S. Greenberg, T. Hudson, M. Levenhagen, A.B. Maccabe, and R. Riesen. (2000). Massively Parallel Computing Using Commodity Components. *Parallel Computing*, 26 (2-3) 243–266.

Burns, L.A., M.C. Barber, S.L. Bird, F.L. Mayer, and A. Suarez. (1990). PIRANHA: Pesticide and Industrial Chemical Risk Analysis and Hazard Assessment. Internal Report, U.S. Environmental Protection Agency, Office of Research and Dev., Athens, Georgia.

Burns, L.A. (1983). Validation of Exposure Models: The Role of Conceptual Verification, Sensitivity Analysis, and Alternative Hypotheses. In *Proc. 6th Symposium - Aquatic Toxicology and Hazard Assessment* (Bishop W.E., Cardwell R.D., Heidolph B.B., eds.), Vol. ASTM STP 802, American Society for Testing and Materials: Philadelphia, Pennsylvania, pp. 255–281.

Burns, L.A. (2001). Probabilistic Aquatic Exposure Assessment for Pesticides - I: Foundations. EPA/600/R-01/071, U.S. Environmental Protection Agency, Office of Research and Development, National Exposure Research Laboratory, Ecosystems Research Division, Athens, Georgia.

Chen, J., and M.B. Beck. (1999). Quality Assurance of Multi-Media Model For Predictive Screening Tasks. EPA/600/R-98-106. U.S. Environmental Protection Agency, Office of Research and Development, Washington, DC.

Cullen, A.C., and H.C. Frey. (1999). *Probabilistic Techniques in Exposure Assessment: A Handbook for Dealing with Variability and Uncertainty in Models and Inputs.* Plenum Press: N.Y., New York.

Funtowicz, S.O., and J.R. Ravetz. (1990). *Uncertainty and Quality in Science for Policy.* Kluwer Acad.: Dordrecht, The Netherlands.

Helton, J.C., and F.J. Davis. (2000). Sampling-Based Methods. In *Sensitivity Analysis* (A. Saltelli, K. Chan, and E.M. Scott, eds.), John Wiley & Sons: West Sussex, England, pp. 101–153.

Kleijnen, J.P.C., and J.C. Helton. (1999). Statistical Analyses of Scatterplots to Identify Important Factors in Large-Scale Simulations, 1: Review and Comparison of Techniques. *Reliability Engineering and System Safety*, 65:147–185.

Marin, C.M., V. Guvanasen, and Z.A. Saleem. (n.d.). The 3MRA Risk Assessment Framework—A Flexible Approach for Performing Multimedia, Multipathway, and Multireceptor Risk Assessments Under Uncertainty. *International Journal of Human and Ecological Risk Assessment* (in press; scheduled for publication December 2003).

Osidele, O.O. (2003). An Integrated Regionalized Sensitivity Analysis and Tree-Structured Density Estimation Methodology. In *Proceedings, International Workshop on Uncertainty, Sensitivity and Parameter Estimation*, Federal Interagency Steering Committee on Multimedia Environmental Modeling, Rockville, Maryland.

Saltelli, A., Chan, K., Scott, E.M.. (2000). *Sensitivity Analysis*. J. Wiley & Sons: West Sussex, England.

U.S. EPA (U.S. Environmental Protection Agency). (2002). Guidelines for Ensuring and Maximizing the Quality, Objectivity, Utility, and Integrity, of Information Disseminated by the Environmental Protection Agency. EPA/260R-02-008. U.S. Environmental Protection Agency, Office of Env. Information, Washington DC. October 2002.

Vose, D. (2000). *Risk Analysis: A Quantitative Guide*, 2nd ed.. J. Wiley & Sons: West Sussex, England.

van der Sluijs, J., P. Kloprogge, J. Risbey, and J. Ravetz, (2003). Toward a Synthesis of Qualitative and Quantitative Uncertainty Assessment: Applications of the Numeral, Unit, Spread, Assessment, Pedigree (NUSAP) System. In *Proceedings, International Workshop on Uncertainty, Sensitivity and Parameter Estimation*, Federal Interagency Steering Committee on Multimedia Environmental Modeling, Rockville, Maryland.

This work was reviewed and approved by EPA.

Ground-Water Parameter Estimation and Uncertainty Applications

Earl Edris

USACOE

A presentation on the section heading topic was given by the speaker identified. No abstract was provided.

Use of Fractional Factorial Design for Sensitivity Studies

Richard Codell

U.S. Nuclear Regulatory Commission
Washington, DC 20555-0001
301-415-8167

RBC@nrc.gov

Factorial design has been used for physical experimentation (Box, 1961) and, more recently, for testing computer codes and models (Andres, 1997). Factorial designs usually sample over a range of each parameter at fixed intervals (e.g., the 5th and 95th percentile for a two-interval design or adding the 50th percentile for a three-interval design). By sampling all parameters in a system in this manner, it is often possible to unambiguously determine the effects of the variations in a parameter and all combinations of parameters. A full-factorial design with M intervals requires MN samples, where N is the number of parameters being examined. However, when the number of parameters exceeds just a few, the number of experiments necessary quickly grows to an unreasonable value.

The NRC staff has been using a suite of techniques to determine parametric sensitivities for a variety of situations in waste management, including low-level and high-level radioactive waste. Such techniques fit into two categories: (1) examining a pool of model results generated from Monte Carlo sampling, and (2) sampling directed by the sensitivity technique itself. The NRC staff has included fractional factorial design to this suite of sensitivity methods for a recent performance assessment of the potential high-level waste repository at Yucca Mountain Nevada (Mohanty, 2002). Fractional factorial methods require far fewer than M^N experiments, but may produce ambiguous sensitivity results. For example, a so-called level-4 design for 330 sampled parameters and two intervals (5th and 95th percentiles of each parameter distribution) required 2,048 samples. Such a level-4 design can yield results for which the main effects of all parameters are distinct from each other and two-way interactions of other parameters, but can be confounded by some three-way and higher interactions of other parameters. Since many of the parameters in the Yucca Mountain case are involved in models for which such interactions are likely, it is important to be able to distinguish true effects of parameters from confounding combinations of higher-order interactions. In many cases, it is possible to use other information generated in the runs to make this determination.

In general, the fractional factorial analysis was conducted in the following steps: (1) develop an initial fractional factorial design for all sampled parameters considering the largest number of runs that reasonably can be handled; (2) from the results of the preliminary screening, perform an analysis of variance (ANOVA) to determine those parameters that appear significant at a specified confidence level (e.g., 95%); (3) screen further the list of statistically significant parameters on the basis of information other than the ANOVA results; and (4) repeat the analyses using a refined set of parameters and higher-resolution designs until results are acceptably unambiguous.

For the example cited, the initial screening employed a level-4 design for 330 parameters at two sampling percentiles (5th and 95th), requiring 2,048 runs. The ANOVA on these results found that there were potentially 100 significant parameters of the 330 at the 95th percent confidence level for the 10,000 year time period of interest. These results were further screened to a list of 37 parameters by observations from other information generated in the simulations; for example, it was possible to eliminate all parameters related to seismic failure of the waste packages by observing independently that none of the waste packages failed by this mechanism and, therefore, that this was a spurious indication caused by higher-order combinations of other parameters.

Using the reduced set of 37 parameters from the initial screening, another level-4 factorial design was set up requiring an additional 2,048 runs. With only 37 parameters, it was possible to observe two- and three-way interactions that were combinations of the main effects and to make inferences about the fourth- and higher-order interactions of those parameters that might be explored by additional factorial designs. This reduced the list to only eight potentially significant parameters, for which a full-factorial design could be constructed with only an additional 256 runs. From the final full-factorial design, it was possible to determine that there were seven significant parameters for the 10,000-year case, and up to at least six-way interactions among these parameters. Results for the 100,000-year time period of interest were generated in the same way, but proceeded more directly to identifying a final list of eight significant parameters because there were more non-zero outputs from the models.

Results from the fractional factorial designs for the 10,000- and 100,000-year time periods of interest were similar to many of the other sensitivity results, although the ranking of the parameters often differed among the various techniques. Monte Carlo results using only the parameters identified by the fractional factorial designs indicates that most of the variance is indeed captured by the identified parameters. We conclude that the fractional factorial method is good at identifying sensitive parameters unambiguously if executed properly. It is also very useful for identifying clearly through ANOVA the interactions among the important parameters. Such interactions were not easily identified by the other sensitivity techniques used.

However, the fractional factorial results are not markedly better than those from other techniques NRC used to identify the most sensitive parameters individually. Among the disadvantages of the fractional factorial technique are (1) it still requires a large number of runs, especially if the number of chosen intervals is greater than 2; (2) it requires a large investment in the analyst's time to screen out possible confounding combinations of other parameters masquerading for the apparently sensitive parameter; (3) the runs required for the sensitivity analyses cannot be used directly to generate the desired output results such as the cumulative distribution of the peak doses; and (4) As used, the results depend on the peak doses generated for each of the runs, whereas the NRC regulations depend on the mean of the distribution of projected doses for 10,000 years after disposal (CFR, 2002).

For this exercise, the NRC staff favored a parameter sensitivity result that combines the results from all of the sensitivity methods. This technique assigns weights to the parameters based on the order they appear in the individual sensitivity methods, and then sums the weights over all methods to determine a final overall ranking. Generating a final result from this list provided the most consistent indication that the sensitive parameters have been identified.

Disclaimer

The NRC staff views expressed herein are preliminary and do not constitute a final judgement or determination of the matters addressed or the acceptability of a license application for a geologic repository at Yucca Mountain.

References

Andres, T.H., 1997, "Sampling and sensitivity analysis for large parameter sets," *J. Statist. Compu. Simul.*, Vol. 57, pp 77–110.

Box, G.E.P., and J.S. Hunter, 2000, "The 2^{k-p} fractional factorial design, Part 1," *Technometrics*, Vol. 42, No. 1, pp 28–47 (Reprinted from *Technometrics*, Vol. 3, 1961)

CFR, 2002, "Disposal of high-level radioactive waste in a geologic repository at Yucca Mountain, Nevada," Office of the Federal Register, *Code of Federal Regulations*, Section 63.113, p 224, U.S. Government Printing Office, January 1, 2002

Mohanty, et al., 2002, "System-level performance assessment of the proposed repository at Yucca Mountain using TPA Version 4.1 code," CNWRA 2002-05, Center for Nuclear Waste Regulatory Analyses, San Antonio, Texas

ISCORS Parameter-Source Catalog

*Anthony B. Wolbarst[1], *, Bruce Biwer[2], Shih-Yew Chen[2], Ralph Cady[3],*
Andrew Campbell[3], Stephen Domotor[4], Philip Egidi[5], Julie Peterson[6], Stuart Walker[1]

* Chair, ISCORS Cleanup Subcommittee – wolbarst.anthony@epa.gov.

[1] U.S. Environmental Protection Agency, Washington, DC 20460;

[2] Argonne National Laboratory, Argonne, Illinois 60439;

[3] U.S. Nuclear Regulatory Commission, Washington, DC 20555-0001;

[4] U.S. Department of Energy, Washington, DC 20402;

[5] Colorado Department of Public Health and Environment, Denver, Colorado 80222;

[6] U.S. Army Corps of Engineers, Omaha, Nebraska 68144.

The efforts of those involved in environmental pathway modeling and risk assessment would be supported by the creation of a national repository of information on parameter values and distributions of known provenance and demonstrated utility. To that end, the Interagency Steering Committee on Radiation Standards (ISCORS) and the Argonne National Laboratory are preparing an online ***Catalog of Existing Sources of Information on Parameters Used in Pathway Modeling for Environmental Cleanup of Sites Contaminated with Radioactivity***. (Member organizations of ISCORS are the U.S. Environmental Protection Agency, the U.S. Nuclear Regulatory Commission, the U.S. Department of Energy, the U.S. Department of Defense, other Federal agencies, and the States of Colorado and Pennsylvania, representing the States.) This ***Parameter-Source Catalog*** is a Web-based, indexed and searchable, readily updateable, and user-friendly compilation of references, compendia, databases, and other sources of information on parameters used in contaminant transport and exposure modeling. Built around a Microsoft® Access® 2000 relational database, it offers subject- and text-search capabilities, provides information on parameter definitions, transport/ exposure pathways, and standard models and codes, and contains a tutorial and frequently asked questions (FAQs) page. The contents are vetted before entry (with acceptance criteria such as publication in a peer-reviewed technical journal, appearance in a formally issued Federal agency report, etc.), which provides some degree of quality assurance. It is anticipated that the database will be filled on an ongoing basis mainly by the users themselves. There is a mechanism by which they can easily submit proposed references to the site managers such that, after they are approved in the quality assurance process, they are automatically entered into the database. The catalog is intended for use by professionals, managers, and others involved or interested in the use of pathway modeling to estimate doses and risks associated with contaminated sites.

7

SESSION 5:

**TOWARD DEVELOPMENT OF A COMMON
SOFTWARE APPLICATION PROGRAMMING
INTERFACE (API) FOR UNCERTAINTY,
SENSITIVITY, AND PARAMETER ESTIMATION
METHODS AND TOOLS**

7.1 Overview and Summary

Editor: Justin E. Babendreier

The final session of the workshop considered the subject of software technology and how it might be better constructed to support those who develop, evaluate, and apply multimedia environmental models. Two invited presenters were featured, along with an extended open discussion on the concept of creating a core "interface level" of programming standards for environmental modeling software.

Discussion was primarily devoted to review of a recently developed experimental Application Programming Interface (API) for uncertainty analysis (UA), sensitivity analysis (SA), and parameter estimation (PE) methods and tools. Designated as the Calibration, Optimization, and Sensitivity and Uncertainty Algorithms API (COSU-API; Appendix A), the API was created through a collaboration of ISCMEM's Software System Design and Implementation Workgroup and the Uncertainty Analysis and Parameter Estimation Workgroup.

The goal of this session was to begin building toward consensus on an adoptable UA/SA/PE API that might one day evolve to meet most, if not all, of the related UA/SA/PE needs of environmental modelers.

Technological Goals of Model Evaluation Science

The previous sessions on UA/SA/PE spoke in many ways to the "science of evaluating models," in theory and in application. Arguably, a desired outcome for model evaluation science is that its existing methods will soon be cast as ergonomic, interoperable, and open source software. Such a technology base, when joined with the right hardware, would provide a critically needed tool set for meeting many of today's modeling challenges. A shared tool set would help us learn about and improve upon models and applications, and would also provide a better understanding of the existing set of evaluation methods and tools available, when and where each is best used, and how we might also improve upon these.

UA/SA/PE help quantify or otherwise qualify the benefits of data quality and quantity. These approaches can identify dominant mechanisms of models, and can also shed light on where advancements may be needed in model construction or the underlying science of models. In view of the public's great interest in broadly acquiring and exploiting such capabilities, an API-based software integration and collaboration effort in UA/SA/PE will hopefully lead to:

- More widespread and appropriate use of more model evaluation tools;

- Greater transparency and confidence in data, models, and model evaluation methods used to support regulatory decision-making; and

- Increased efficiency and accuracy in the identification of key parameters and processes that dominate model output behavior.

With a growing reliance on models to support increasingly complex decisions, an integrated UA/SA/PE tool box should also help modelers keep up with new quality assurance guidance (EPA, 2002, 2003).

What Is An API and What Are Its Benefits?

An API is a standard set (or library) of functions, variables, and constants that software developers can leverage to achieve a high level of functionality and interaction with other software programs. An API is formally defined as a set of software calls and routines that can be referenced by an application program in order to access supporting network services (ANSI, 2001). An API allows software developers to easily incorporate API-compliant software without having to know the details of how the software's functionality is implemented—hence, the term "interface."

An API can be especially useful, and is increasingly essential, when the works of many software developers are to be integrated across many institutional boundaries. Development of a flexible, yet useful set of standards appears to be an imminently logical step for the Federal research community. A UA/SA/PE API would, together with other APIs (e.g., I/O, GIS, visualization, etc.), deliver a greatly enhanced ability for stakeholders and regulators to leverage environmental modeling software products across agencies and other institutions. A widely adopted UA/SA/PE API would be expected to appreciably accelerate achievement of the technological goals of model evaluation science.

API Session Outline

The API session created an opportunity for direct technologist-to-scientist discussions on the subject of creating modern (and to some degree object-oriented) standards for UA/SA/PE software tool developers. Environmental software engineers exchanged ideas with the many workshop participants who develop, apply, and build UA/SA/PE methods and tools. As a group, the workshop participants were expected to encompass a broad range of software programming skills and levels of familiarity with session topics.

The first presentation on software technology focused a multi-agency perspective of modeling system "framework" development. It was given by Gerry Laniak of USEPA's Office of Research and Development who serves as co-chair of a companion ISCMEM workgroup that focuses on software technology development for science-based modeling. This initial discussion formed a foundation for the subsquent discussion by Karl Castleton of PNL-DOE, of the same workgroup, who presented the experimental multi-agency COSU-API under development for model evaluation methods and tools. The session proceeded by first introducing key concepts in framework technology, next presenting the draft COSU-API in somewhat lay terms, and finally seeking open discussion from the audience on how well the draft API supports the goals and needs of the UA/SA/PE software tool development community.

7.1.1 Discussion Questions

The API Development Team posed the following questions to facilitate the discussion:

1. Why is the UA/SA/PE API important to non-programmers?

2. How important is nesting of operations?

3. Are tables sufficient for data exchange between UA/SA/PE components?

4. Where are the logical connections between UA/SA/PE components (i.e., where are tables produced and consumed)?

5. How should UA/SA/PE components be run?

7.1.2 Discussion Summary

7.1.2.1 Key Concepts in Framework Development

Creating a setting for introduction of the COSU-API, the opening presentation and discussion on frameworks attempted to outline answers to the following framework-related questions:

- What is a "modeling framework"?
- What are the attributes of a modeling framework?
- What can a framework do for UA/SA/PE method development and application?
- What are some issues that remain to be resolved with respect to frameworks?

The general notion of a modeling framework as a "system infrastructure" was introduced, analogous, for example, to the software that glues Microsoft® Office components together. Modeling framework "infrastructure" components were generically described as the software tools that facilitate the development, organization, and execution of integrated solutions to modern environmental assessments. A modeling framework's primary function was depicted as facilitating the integration of the science behind these assessments, in the form of models and databases (and various tools).

Elements of a Framework

Typically, a modeling framework encompasses the following elements:

- Science-based models;
- Environmental databases;
- User interface(s);
- System-level execution management;
- Methods for managing input and output (I/O) data within the framework;
- Geographic Information System (GIS)-based data access, organization, viewing, and analysis;
- Model evaluation tools (e.g., Monte Carlo simulation, COSU-API, etc.); and
- Other data analysis, visualization, and distributed computing tools.

These elements are drawn out in alternate organization in Figure 7-1. Core infrastructure framework elements would be those other than models, databases, and model- or module-specific user interfaces.

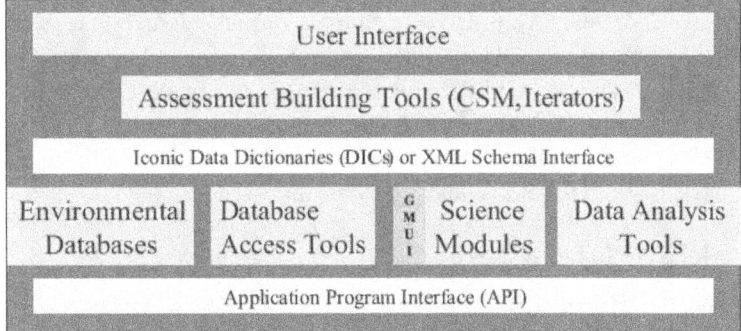

Figure 7-1: Elements of a Modeling Framework

157

There are several (if not a plethora) of modeling frameworks currently in existence and in some form of use by Federal agencies participating in the workshop (e.g., FRAMES, OMS, MIMS, GMS, DIAS, GoldSim, BASINS, etc.). Each of these frameworks essentially constitutes a different set of APIs, with varying levels of sophistication, standardization, commonality, and focus (e.g., approaches for execution management, I/O, GIS, UA/SA/PE, statistical analysis, graphical analysis, etc.). In characterizing the state of modeling, there are many active software system development approaches, and the associated framework technologies share relatively few (if any) common standards across development groups. Extending this, there are also examples of non-standard, intra-framework I/O management, where the framework may not require use of a shared I/O API.

Key concepts that define frameworks were described and included (1) inter-component data transfer; (2) plug-n-play capability; (3) meta-data; and (4) APIs. The discussion on APIs considered examples at the environmental science and computer science levels.

UA/SA/PE and the Role of Frameworks

Reasons or selling points suggested for developing UA/SA/PE tools under a common API for use within a variety of framework environments included the following notions:

- UA/SA/PE methods are applicable across a broad modeling and assessment domain.

- UA/SA/PE methods would receive wider use and, thus, more feedback to the developer.

- Frameworks spawn collaboration.

- Frameworks open up new worlds of modeling experimentation (e.g., allowing for the direct comparison of models, UA/SA/PE methods).

- Frameworks do not constrain the expression of science, they expand it.

In summarizing, development of a common UA/SA/PE API is, on some level, equivalent to creating a framework-independent approach. A key term introduced was "interoperable"; that is, components that will operate equally well in all frameworks. One can imagine for a moment how simple collaboration might be if the only differences between agency frameworks were the underlying science and data they imported, and the custom user interfaces they created. While reaching a common set of "core element" APIs across all frameworks is (likely) too ideal to reach under any circumstances, there are several experiments underway for broad-scope, interoperable concepts like GIS, UA/SA/PE, and I/O.

A specific point offered from the audience included a notion about how the ISCMEM agencies might market this concept, perhaps to the Office of Management and Budget (OMB). There were many comments also made about how "this time has come," and how other similar efforts are underway. Comments were also offered from various international guests that they are interested in collaborating on these concepts. Harmon/IT discussed how their common framework sounded as if it mimicked EPA's FRAMES elements on core, but not specifically for the UA/SA/PE API aspects. The issue of software programming language independence arose on discussion of APIs. An example of the FRAMES I/O API and its viability for compiling in four languages (Java, C++, FORTRAN, and VB) was discussed.

Some Framework Issues to be Resolved

Notably touched on in this session's discussions was the fundamental concept of I/O management across and within frameworks, and associated approaches for managing the iconic structure of environmental assessments (e.g., data dictionaries, XML-based schema). Issues to be resolved included discussion on units on values (or not), bounding values between transfers (or not), arrays of values versus true object-oriented structures, managing error and warnings, and meta-data about models (or implied by use). An open question was left to the audience by the presenter: "If we are all building our own framework, how much time are we spending on development of science and data, as opposed to the core infrastructure?"

7.1.2.2 Conceptual Overview of the Multi-Agency COSU-API

For the second presentation, the multi-agency COSU-API opened with an initial summary of their answers to the five lead discussion questions raised in Section 7.1.1. The technologists generally indicated that nesting of UA/SA/PE components is centrally important, and the concept of "table" appears to serve well as the primary data communication mechanism between UA/SA/PE components. A "table," conceptualized as a spreadsheet page, was described as a simple row-column organization of variables (each column) and iteration values (each row). They mentioned that some alternative concepts of bounding were perhaps needed in structuring tables (e.g., for representing other than a default column-variable, row-iteration assumption). It was also remarked upon that many UA/SA/PE components already tend to have this "table" concept built-in.

A discussion in finding and identifying the natural connections between UA/SA/PE components that produce or consume a "table" was eventually taken up, as was the concept of why it would be good to have UA/SA/PE components run in a standard manner. Further elaboration on and assessment of these answers seeded by the technologists were preceded by a discussion of who might be interested in creating a UA/SA/PE API and why. The discussion on API stakeholders covered the perspectives of managers, scientists, programmers, and various computer science experts present at the workshop.

UA/SA/PE Components

A UA/SA/PE component was defined as a piece of software that contains algorithms that support the process of producing UA/SA/PE results. These were noted to be different from science models typically thought of in modeling frameworks, and each would be, in some way, smaller than the entire framework subsystem that performs UA/SA/PE. Further, API-bound components were characterized as being those that are "reusable" across modeling exercises and across computer languages ("reusability").

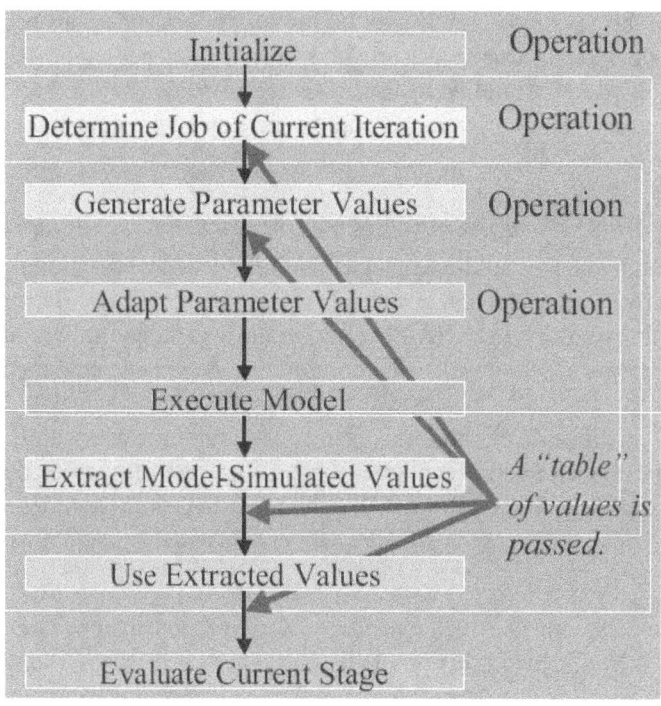

Figure 7-2. Jupiter Example Conceptual Layout

Covering the general notion of "tables" and identifying natural connections between UA/SA/PE components, Figure 7-2 shows a thematic example of Jupiter's conceptual layout, which was used at several points during the presentation and discussion. Currently under development by USGS and EPA, the Jupiter technology (Joint Universal Parameter IdenTification and Evaluation of Reliability - Section 3.2.?) is one of the initial applications of the COSU-API. Jupiter will combine many of the existing PEST (Section 3.2.?) and UCODE, (Section 3.2.?) functionalities.

A few descriptors for UA/SA/PE components given included, "samplers" (a producer of tables), "summarizers," "Monte Carlo," and "data visualization" (a consumer only of a table).

From Appendix A, interface, class, and exception summaries for the COSU-API are given in Table 7-1.

**Table 7-1: Interface, Class, and Exception Summaries for the COSU-API
I/O, Execution Management, and Frameworks**

Interface Summary

ComplexTable	A table that can hold any type of information.
ComplexTransformation	An interface for a transformation that is applied to a ComplexTable and produces a ComplexTable.
DoubleTable	This is the basic table for accessing floating point numbers.
DoubleTransformation	An interface for a transformation that is applied to a DoubleTable and produces a DoubleTable.
Executer	An Executer provides Operation execution queuing and control.
Operation	Operation represents a significant computation.
SelfDescribingOperation	An interface that allows Operations to describe themselves.
SimpleTable	This type of table can store floating point numbers (doubles), booleans, integers, and strings.
SimpleTransformation	An interface for a transformation that is applied to a SimpleTable and produces a SimpleTable.

Class Summary

ByReferenceBoolean	This provides a way for methods to return a boolean value through their argument list.
Column	This class describes a column in a data table.
RowException	Information about an exception related to a row in a table.

Exception Summary

TableException	An exception thrown when one of the semantics of operations on tables has been violated (e.g., close a table that has not been opened, access a table that has not been opened, access a value that has not been set).

The "Double," "Simple," and "Complex" tables and associated "Transformations" handled in the COSU-API represent the primary data communication mechanisms between UA/SA/PE components. In addition, "Operation" and "Executer" interfaces are also present. "Operations" provide a standard execution mechanism that consumes and produces tables. One can think of "Transformations" as simple "Operations." An "Executer" provides "Operation" execution queuing and control. Both were added to standardize and facilitate implementation and integration of UA/SA/PE components. Defining execution management (EM) tasks, "Operations" and "Executions" capture, in some sense, a separate EM-API (e.g., run, restart, error/exception handling, queuing, etc.). One concept suggested was that the "Executer" and "Operation" interfaces might serve well as an initial EM-API for all modeling framework technologies. Discussion on whether the COSU-API is adequate will hopefully build toward a consensus in the modeling community on standards for both EM and UA/SA/PE functionalities.

Like EM, due to a lack of widely adopted standards and the desire to facilitate application development, the COSU-API also implements some basic I/O functionality for "Tables" (e.g., handling scalars and one-dimensional data types, minimal meta-data, etc.). As a reference point for interpreting themes in this session, one can characterize Jupiter as a specific framework implementation of the UA/SA/PE, EM, and I/O components of the COSU-API. In addition to having its own unique interface(s) and data analysis tools, Jupiter also expands upon the minimal COSU I/O functionality, for example, by further describing information in "Table" columns, wrapping models, specifying file formats, expanding meta-data, and so forth.

COSU-API Functionality

The COSU-API was intended by its designers to be a simple, easily implemented API. A basic rule of thumb offered on identifying candidate UA/SA/PE components was that if it tends to take a table and produce a table of results, a UA/SA/PE component should probably conform to the "Operation" COSU-API functions for the sake of reusability. An "Operation" was characterized as only being concerned with computing its results based on the input table(s) it is given. How operations are nested can be (1) fixed (as in the JUPITER diagram); (2) free-form diagrammed, as in FRAMES and MIMS; or (3) simply reused in an application such as DIAS and OMS. In terms of where and when to use the COSU-API, several rhetorical questions were posed:

- Have you reused this functionality before?

- Have you cut and pasted code into another program?

- Does your routine tend to produce or consume "Tables"?

- Is a "Table" a natural form of the information?

A key point about execution, the "Executer" interface allows for distinguishing between where a component is executed and where it is invoked (supporting execution of parallel operations). While emphasizing the presence of the "Executer" interface in the COSU-API, the presentation primarily focused on specific examples of functions found in "Operations" and "Tables."

Implementing Operations Across Languages

To underscore the flexibility in implementing the COSU-API, some specific examples of interface functions were given. Emphasizing the COSU-API's ability to be implemented across object-oriented and legacy programming styles, linkage approaches for Java/C++ and FORTRAN/C were offered and distinguished. Table 7-2 shows operation function structures discussed for each style.

The technologists explained that in going from FORTRAN/C to Java/C++ based interfaces, one would compile the code as a dynamic link library (DLL) or shared object (SO) that contains the seven basic "Operation" functions captured in Table 7-2, where source code could be delivered as well. Java or another object-oriented (OO) language would wrap the specific use (i.e., instance) of an operation to an object, and multiple instances of the operation could then be used by the OO language. They also pointed out that in going from Java/C++ to FORTRAN/C, a single instance of the Java/C++ object would be wrapped in a DLL or SO. A FORTRAN module (or C header) would be created that would allow the program to call the appropriate functions. The COSU-API would even support rudimentary approaches still in practice (for example, the legacy programming approach of using integers as handles, analogous to file numbers).

162

Table 7-2: Operations Structure for OO and Legacy Programming Styles

Java/C++ Interface *(OO)*	FORTRAN/C Interface *(Legacy)*
• Boolean canRestart()	• subroutine myOP_cleanup(integer OpId)
• void cleanup()	

• DoubleTable restart(
    DoubleTable input,
    DoubleTable partialResult,
    ByReferenceBoolean complete)
	• integer function myOP_restart(
	    integer OpId
	    integer input
	    integer,partial
	    logical complete)

• DoubleTable run(
    DoubleTable input,
    ByReferenceBoolean complete)
	• integer function myOP_run(
	    integer OpId
	    integer input
	    logical complete)

• void setup()	• subroutine myOP_setup(integer OpId)
• void stop()	• subroutine myOP_stop(integer OpId)
• Boolean supportsParallelRuns()	• logical function
• Boolean myOP_canRestart()	• myOP_supportsParallelRuns(integer OpId)

Table Functions

The COSU-API offers three separate interfaces, one each for double, simple, and complex "Tables," where these extend from each other in this order. There are also three separate "transformation" interfaces, one for each "Table" type. Examples of DoubleTable functions found in the COSU-API were explained and are restated in Table 7-3.

Extending DoubleTable, SimpleTable can store and retrieve strings, integers, logicals (booleans) and doubles. Further, ComplexTable extends (i.e., is derived from) SimpleTable and can handle any data type (assuming one is using an object-oriented programming language). DoubleTransformation takes a DoubleTable and produces a DoubleTable, where the transformation can be used, for example, to make subsets or encapsulate summarization techniques. SimpleTransformation and ComplexTransformation can do the same for their associated "Table" types.

Revisiting the concept of extended meta-data not handled in the COSU-API, more information about what is in the actual columns of "Tables" would be made available through the Column "class." This would be managed through additional I/O API functionality provided by the user during implementation of the COSU-API for their specific applications.

Table 7-3: DoubleTable Functions

- subroutine close(
    Integer TableId)

- integer function findColumnByName(
    integer TableId
    character(*) colName)

- integer function findRowByName(
    integer TableId
    character(*) rowName)

- integer function getColumnCount(
    integer TableId)

- character(*) function getColumnName(
    integer TableId
    integer columnIndex)

- double function getDoubleAt(
    integer Tableid
    integer rowIndex
    integer columnIndex)

- double(*) function getDoubles(
    integer TableId
    integer rowIndex)

- integer function getRowCount(
    integer TableId)

- character(*) function getRowName(
    integer TableId
    integer rowIndex)

- logical function isCellEditable(
    integer TableId
    integer rowIndex
    integer columnIndex)

- logical function isValueAvailable(
    integer TableId
    integer rowIndex
    integer columnIndex)

- subroutine open(
    integer TableId)

- subroutine setDoubleAt(
    integer TableId
    integer rowIndex
    integer columnIndex
    double aValue)

- subroutine setDoubles(
    integer TableId
    integer rowIndex
    double[] values)

- subroutine setRowName(
    integer TableId
    integer rowIndex
    character(*) name)

- subroutine waitForDataAvailable(
    integer TableId
    integer rowIndex
    integer columnIndex)

7.1.2.3 Open Audience Discussion on the COSU-API

Summarizing some of the major exchanges between the technologists' presentations and participants in the workshop, several notable comments were offered. As a starting point, in the open discussion period, the presenter of the COSU-API led with a question on "Operations." Specifically, are "initialize" and "determine job" actually operations, and do they need to be addressed in the COSU-API. This was followed by some discussion on the concept of language independence and perspectives on "simple" and "double" tables, and their believed fit to the participants' existing UA/SA/PE tools. Rows versus columns as variables vs iterations was further reviewed. API designers indicated they may eventually reconsider this arrangement, although no specific alternative was arrived at in discussions.

On aspects of execution management, a discussion was pursued on how warnings from legacy code might be transferred at the higher (system) level for user access. Some consider this a module-level responsibility/activity. The conversation distinguished between warnings and errors, where it was noted that EM of the COSU-API provides error handling. One commentator noted that for FORTRAN, like it or not, it is a typical choice by many developers, and it would be nice if a system-level "warning" handling capability could be built into the next version (i.e., of the COSU-API).

One participant described a lack of seeing what the unique aspect of UA/SA actually is. This led directly to the review of the concept of "Tables." An example was also given on how general I/O would need several techniques to handle data. The example considered the concept of several models, with gradation in output data storage needs, amenable to a single Latin Hypercube Sampling (LHS) run for UA. Consider that one code stores lots of data, the next needs less. This was followed by discussion of existing I/O API approaches found in EPA's MIMS (a minimalist approach) versus the FRAMES development concept which encourages use of a standardized I/O API for all framework components. Like the COSU-API, there is wide flexibility in MIMS with minimal standards on I/O functionality. In FRAMES, registered components that comply with the FRAMES I/O API are plug-and-play, so to speak, with core framework components and the variety of other API-linked components (e.g., models, databases, post-processors, etc.). A FRAMES philosophy is that I/O standardization is key to integration and quality assurance.

A discussion on nesting was taken up where it was indicated that the presentation example given was viable. One point made was the idea that UCODE can be used by UCODE. A question arose on what is the percent effort of facilitating nesting. One technologist indicated that requirements may become difficult when we set standards for "table." On pursuing further what was meant, it was commented that the specificity of saying you want a specific thing is sacrificed therefore, an extra level of documentation for how different people use this "table" idea is needed. One participant said the shocking thing about this (COSU-API) is that "distribution class" doesn't roam around; the API appears to be very low-level.

A comment made from the SA point of view was that it was most suitable for sampling-based approaches. One participant indicated that SA methods may use the entire model as part of the entire model (i.e., first sample, run model, have model output, that output is analyzed to come up with S parameters). The conversation led back to the idea that "table" can store this information. One software engineer asked about how relational or normalized data will be handled. For example, how will large tables be handled. It was indicated, in response, that the COSU-API hasn't said how these will be stored; the API is flexible here. Another commentator asked what about inherent capabilities in some data systems that can already do statistical analysis. It was discussed that a "transform hook" could possible assign this task to the host database. In database discussions, it was indicated that for sparse memory or storage issues, these are really to be handled by the user. Some major points offered on the COSU-API were that:

- The COSU-API does not dictate how to store or handle data (i.e., extended I/O).

- The user decides how to make their "tables" complete.

- The COSU-API is simply saying what actions need to be fulfilled, not how they are done.

Finally, an example was given on why Microsoft® went to the use of dynamic link libraries (DLLs). This was followed by discussion of the concept of DLL versus executable, and the question of whether we want source code as basis.

7.1.3 Application Issues

The COSU-API is currently being used in some initial UA/SA/PE applications [for example, as a basis for software code development in the Jupiter project (see Section 3.7)]. A few other COSU-API application efforts are also underway, including work associated with OMS, MIMS, and FRAMES.

On a conclusive note about the session's primary theme, the multi-agency API design team reiterated that they thought that the COSU-API was a proposal worth considering. An overview comment was also made that while the experimental COSU-API is not substantially implemented anywhere as yet, on functionality, it is essentially already implemented in many places.

At this point, the draft COSU-API documentation is being made freely available, although the class files and source code are not currently easily accessed outside of ISCMEM. The intention is that various ISCMEM members will first directly evaluate the adequacy of the COSU-API in the noted implementations underway. This will allow for a trial phase before attempting to support the API publicly. In this sense, "lessons learned" is a developing story. Interested parties were invited to contact the ISCMEM workgroups directly to submit comments, ideas, or proposals involving the COSU-API.

The COSU-API and Modeling Frameworks

Of particular interest, is the related role of modern modeling frameworks as an application medium for conducting and investigating model evaluation science. Consider the analogy that frameworks are the "office buildings" that house models and data, where information easily flows from one room to another, and from one floor to another. Consider also that model evaluation tools, similar to all pre- and post-processors that act upon data, are really just models that act upon models and databases. In frameworks, models, databases, and tools are analogous in interoperable communication and the operations performed upon them. Core framework elements, including generic tools, would be found in the basement, so to speak. In appearance and (likely) functionality, each building or framework is characterized by a unique front door, a "boiler room" with an exoteric or esoteric feel to it, and possibly a different set of models and databases. Framework applications would be the use of these buildings for specific assessments.

The proposed premise of the COSU-API is, reasonably, that it would be beneficial to create a "living" tool once that many frameworks can accept (and improve upon), and through which such tools could be applied to many models and databases (i.e., creating a more exoteric boiler room). Because model evaluation tools consume and produce information in a similar, structured context (e.g., "tables," nested operations, etc.), this commonality between UA/SA/PE tools defines the underlying advantage and form of an associated API. If sharing and leveraging the best available model evaluation science is a goal, the strategic investment of accepting standard programming practices for UA/SA/PE tools offers a solution. It would need to be a widely adopted, well-supported API that evolves over time to meet its users' needs.

7.1.4 Lessons Learned

Given the relatively immature status of the COSU-API, lessons learned are posed here instead as a set of more detailed questions regarding the utility of the draft COSU-API that, in part, remain to be answered. These same questions were provided as guides during the open discussion period, and will need to be more formally deliberated upon.

The added detail of the questions posed below further explicates the information likely needed for a thoughtful review of the COSU-API. As the COSU-API is put through its initial paces, it will hopefully move forward in building consensus on a formal, adoptable standard. The API would, by its nature, be expected to evolve over time as we learn more about the collective set of model evaluation methods and tools, assessment needs, and technological advances in computer science.

On Nesting of UA/SA/PE Components

- Is general nesting achievable or even desirable?

- Does it over-complicate the implementation of individual components?

- Does it add to the overall capability?

- Is this too simplistic of a viewpoint?

- If so, what aspect is too simplistic?

On Tables as a Primary Mechanism Between UA/SA/PE Components

- Is this too simplistic?

- Maybe for the whole framework, but what if just for UA/SA/PE components?

- Is this not simple enough?

- What mechanism would be better and achievable by the group (implies that multiple programming languages, not all object oriented, would need to be supported)?

- Is this concept separable from the data storage mechanism?

On Finding the Natural Connections Between UA/SA/PE Components

- Are their standardized locations of connections?

- Do these connection points get us most of the reusability?

- Is the "table" restricting us from connections that would be more natural?

On Running UA/SA/PE Components in a Standard Way

- What is the deliverable for a component?

- Currently people deliver executables. Are DLLs possible?

- How independent can the running of components be?

- How important is parallel execution?

- How are tables handed components?

7.1.5 Research Needs

Identifying next steps leading to acceptance, modification, or formalization of the COSU-API, a few summary comments were made on the question of research needs. The API designers stated a specific intention to leave the workshop, go back, and reflect upon the sentiments and questions raised. There was general consensus that, as a group, we should next apply the draft API in some initial test applications, and determine its adequacy and fit. In addition to Jupiter, MIMS, and FRAMES, an additional example of the GEOLEM project was given where USGS is also working with the COSU-API to put some of these capabilities together (e.g., OptTool). GEOLEM is another ISCMEM API project underway for standardizing some core capabilities in geographical information systems (GIS). It was noted that EPA's MIMS project has already started to implement some of the COSU-API concepts.

A key question offered by the technologists was to ask developers to "consider what you are doing in the areas of UA/SA/PE and look at the COSU-API." Stressing the interoperable perspective of the COSU-API, the prognosis is that one will gain the ability to more easily share UA/SA/PE components and actual results, regardless if you are working within modern modeling frameworks like MIMS, OMS, FRAMES, and so forth.

As further evidence of consolidation underway in software design standards, there was also mentioned a desire to possibly pursue evaluation of the "R Project for Statistical Computing" (i.e., the R API as an alternative to SAS, NCSS, etc.). This would be seen as a complementary API for delivering interoperable data analysis and visualization tools. Based on Bell Laboratories' "S" language, "R" is an integrated suite of freeware software facilities for data manipulation, statistical calculation and "publication ready" graphical display (www.r-project.org). R is considered highly "extensible." On the question of peer-review of the R API, it was mentioned that R's graphics were thought ok, but its statistical routines may not be significantly peer-reviewed as yet (e.g., perhaps not yet having similar levels of acceptance as SAS for use in publishing data analysis in literature).

The rather large issue of proprietary code was finally revisited in the context of something that needs to be resolved. Many varied opinions were offered on the subject. It was mentioned that EPA's Council on Regulatory Environmental Modeling (CREM) would attempt to address some of these third party issues in new guidance developed, which is still undergoing final peer-review stages (EPA, 2003).

7.1.6 Conclusions

Multimedia environmental modeling could benefit considerably from a robust software language structure that will lead to the ease of anyone to readily and efficiently integrate UA/SA/PE tools with models and data. For multiple agencies who are expending significant resources on core science research and software development, there is obvious potential benefit to be realized from our tools "speaking the same language" within agencies, across agencies, and across time. The idea is to build and enhance the core infrastructure of modeling frameworks once, for all to use, and then concentrate on science development.

While not specifically addressing extended I/O standards (which ultimately need to be addressed for the larger grouping of all models, databases, and tools), the model evaluation API presented here (the COSU-API) sets forth a potentially useful scheme for organizing, describing, and executing model evaluation tasking (e.g , simulation experimentation, pre/post-processing, nesting of operations, UA/SA/PE, parallel processing, etc.). Key questions to answer are will the draft API be flexible enough, and is it adequate? This should be asked for each of the UA/SA/PE, EM, and I/O functionality sets found in the COSU-API.

Typical of most existing modeling frameworks (FRAMES, OMS, Jupiter, etc.), each, in its own way, implements some approach (or API) for I/O, EM, and iconic data schema. As a group, there remains a notable lack of inter-agency and intra-agency consensus on I/O and EM standards. With the COSU-API as a starting point for discussion, it is possible, at least, that standardization can continue to be further addressed by ISCMEM's Software System Design and Implementation Workgroup. This may likely proceed for now in the order of addressing UA/SA/PE, EM, GIS, data analysis and visualization, and I/O.

With our growing reliance on model outputs to support increasingly complex regulatory decisions, a working assumption should be that an integrated UA/SA/PE tool box would be best sooner, not later. Establishing a widely adopted multi-agency API, clearly, is easier said than done. One participant remarked on the worry of a proliferation of APIs. As an interesting point of argument, alternatively, we might consider that this, in fact, has already occurred, and will continue to define the status quo for environmental modeling until remedied through efforts like the COSU-API.

7.1.6 References

American National Standards Institute (2001). "Telecom Glossary 2000." ANS T1.523-2001.

EPA (2002). "Guidelines for Ensuring and Maximizing the Quality, Objectivity, Utility, and Integrity of Information Disseminated by the Environmental Protection Agency." U.S. Environmental Protection Agency, Office of Environmental Information. EPA/260R-02-008. http://www.epa.gov/quality/informationguidelines/index.html.

EPA (2003). "Draft Guidance on the Development, Evaluation, and Application of Regulatory Environmental Models." U.S. Environmental Protection Agency, Office of Research and Development, Office of Science Policy, Council for Regulatory Environmental Modeling (CREM). http://cfpub.epa.gov/crem/cremlib.cfm.

An Overview of the Uncertainty Analysis, Sensitivity Analysis, and Parameter Estimation (UA/SA/PE) API and How To Implement It

Karl Castleton[1], Steve Fine[2], Ned Banta[3], Mary Hill[4], Steve Markstrom[5], George Leavesley[6], and Justin Babendreier[7]

[1] PNNL, DOE, Operated by Battelle Memorial Institute, Richland, Washington, USA, karl.castleton@pnl.gov

[2] U.S. Environmental Protection Agency, Research Triangle Park, North Carolina, USA, fine.steven@epa.gov

[3] U.S. Geological Survey, Lakewood, Colorado, USA, erbanta@usgs.gov

[4] U.S. Geological Survey, Boulder, Colorado, USA, mchill@usgs.gov

[5] U.S. Geological Survey, Boulder, Colorado, USA, markstro@usgs.gov

[6] U.S. Geological Survey, Boulder, Colorado, USA, george@usgs.gov

[7] U.S. Environmental Protection Agency, Athens, Georgia, USA, babendreier.justin@epamail.epa.gov

The Application Programming Interface (API) for Uncertainty Analysis, Sensitivity Analysis, and Parameter Estimation (UA/SA/PE API) [also known as Calibration, Optimization and Sensitivity and Uncertainty (CUSO)] was developed in a joint effort between several members of both the Framework Software Workgroup and the Uncertainty and Parameter Estimation Workgroup of the Federal Interagency Steering Committee on Multimedia Environmental Modeling (ISCMEM). The primary purpose for its undertaking, the development of the current draft UA/SA/PE API presented here, attempts to initiate discussion and increase cooperation among the various Federal agencies in moving toward a common software programming approach for the future development of sharable tools and methods for conducting uncertainty analysis, sensitivity analysis, and parameter estimation.

Pivoting from the previous discussion on the related role of environmental modeling frameworks, the UA/SA/PE problem set represents a potentially fruitful area of common agreement among Federal researchers in standardizing software technology development. The vision of a final API, still to be agreed upon, is to eventually allow all model developers, model users, regulators, and stakeholders to more readily benefit from each other's efforts, accomplishments, and insights into these important areas of model evaluation. Such cooperation is envisioned to greatly accelerate the Federal agencies' collective capability over the next decade to more objectively compare the utility of various available methods, tools, and techniques, and to better understand their strengths and weaknesses in solving a wide range of model investigation questions currently faced by the community.

The team's API development strategy sought first to initially produce a relatively flexible, lightweight design in order to allow for inclusion of new approaches, while supporting advanced capabilities and work across multiple operating system platforms, computer programming languages, and modeling frameworks. The API is informally introduced here, setting the stage for a more in-depth discussion with workshop participants as to the positive attributes and potential shortcomings of the API in meeting the multiple needs of the diverse group of researchers in this area. The formal, draft API specification is available upon request, and will also be published within the workshop proceedings, along with comments and concerns raised in the discussion. The purpose of this presentation, and discussions to follow, is to introduce the API, and to solicit critically important input from the modeling community.

The focus of this introductory presentation will center on the following themes:

(1) A component should be able to run within other components. For example, a Monte Carlo tool should allow for additional Monte Carlo stages that are operated by other tools.

(2) A table (e.g., a spreadsheet page) structure appears adequate for communication between these tools. A common approach for transferring data between components is to use a structure akin to a page in a spreadsheet.

(3) There are points between the components of existing toolsets that the API should be injected between. These toolsets may need to provide the ability to produce or consume these tables to take full advantage of the reusability that the API provides.

(4) The invocation of the components needs to be standardized. A single, simple method for invoking the components needs to be agreed upon and followed.

Background information will be provided on each theme that places the associated conclusion in perspective. The four themes above will be illustrated where examples of API implementations will be given using different programming languages. The audience will then have an opportunity to comment on the group's conclusions during the open, participative discussion that follows this presentation.

APPENDICES

APPENDIX A

Calibration, Optimization, and Sensitivity and Uncertainty Algorithms Application Programming Interface (COSU-API)

Lead Developers:
Steve Fine, Karl Castleton

Contributors:
Ned Banta, Mary Hill, Steve Markstrom,
George Leavesley, Justin Babendreier

Editor:

Justin Babendreier
U.S. Environmental Protection Agency
Office of Research and Development

CONTENTS

1.0 Calibration, Optimization, and Sensitivity and Uncertainty Algorithms API

The Application Programming Interface (API) for Uncertainty Analysis, Sensitivity Analysis, and Parameter Estimation (UA/SA/PE API) tool development, referred to here as the Calibration, Optimization, and Sensitivity and Uncertainty Algorithms API (COSU-API), was initially developed in a joint effort between several members of both the Framework Software Workgroup and the Uncertainty and Parameter Estimation Workgroup of the Federal Interagency Steering Committee on Multimedia Environmental Modeling (ISCMEM).

The draft COSU-API (Version: June, 2003), presented formally in this document, attempts to initiate discussion and increase cooperation among the various Federal agencies in moving toward a common software programming approach for the future development of sharable tools and methods for conducting uncertainty analysis, sensitivity analysis, and parameter estimation. Overview elements of the COSU-API were initially presented and discussed among participants attending the August 2003 ISCMEM International Workshop on Uncertainty Analysis, Sensitivity Analysis, and Parameter Estimation.

A complete set of documentation for the COSU-API is available to the public and may be found at the web site http://mepas.pnl.gov/Wiki/page.jsp?website=UASAPE. The download site also includes instructions for submitting comments and contributions to further enhance the API.

1.1 HTML-Based Documentation for the COSU-API

The original HTML-based electronic documentation for the COSU-API provides pages corresponding to the items in a master navigation bar, described as follows.

Package Class **Tree Deprecated Index** `Help`

PREV NEXT   FRAMES   NO FRAMES   All Classes

The proceedings format presented here has attempted to capture the original electronic documentation format, with minor editing to allow for printed document section enumeration and layout.

1.2 How This API Document Is Organized

Elements of the COSU-API documentation provided in this report are organized along the following descriptions and are geared for software engineers and UA/SA/PE technologists.

Package

An API package generally has a documentation page that contains a list of its classes and interfaces, with a summary for each. This page can contain four categories:

- Interfaces (italic)
- Classes
- Exceptions
- Errors

Class/Interface

Each class, interface, nested class, and nested interface has its own separate page. Each of these pages has three sections consisting of a class/interface description, summary tables, and detailed member descriptions:

- Class inheritance diagram
- Direct Subclasses
- All Known Subinterfaces
- All Known Implementing Classes
- Class/Interface Declaration
- Class/Interface Description
- Nested Class Summary
- Field Summary
- Constructor Summary
- Method Summary
- Field Detail
- Constructor Detail
- Method Detail

Each summary entry contains the first sentence from the detailed description for that item. The summary entries are alphabetical, while the detailed descriptions are in the order in which they appear in the source code. This preserves the logical groupings established by the programmer.

Tree (Class Hierarchy)

There is a Class Hierarchy page for the package, plus a hierarchy for the package. Each hierarchy page contains a list of classes and a list of interfaces. The classes are organized by inheritance structure starting with *java.lang.Object*. The interfaces do not inherit from *java.lang.Object*.

- When viewing the Overview page, clicking on "Tree" displays the hierarchy of the package.
- When viewing a particular package, class, or interface page, clicking on "Tree" displays the hierarchy for only that package.

Deprecated API

The Deprecated API page lists all of the API, that have been deprecated. A deprecated API is not recommended for use, generally due to improvements, and a replacement API is usually given. Deprecated APIs may be removed in future implementations.

Index

The Index contains an alphabetic list of all classes, interfaces, constructors, methods, and fields.

Prev/Next

These links take you to the next or previous class, interface, package, or related page.

Frames/No Frames

These links show and hide the HTML frames. All pages are available with or without frames.

Serialized Form

Each serializable or externalizable class has a description of its serialization fields and methods. This information is of interest to re-implementors, not to developers using the API. While there is no link in the navigation bar, you can get to this information by going to any serialized class and clicking "Serialized Form" in the "See also" section of the class description.

Help

The COSU-API help file is based upon API documentation generated using the standard doclet.

2.0 COSU-API Package Summary

An Application Programming Interface (API) for calibration, optimization, and sensitivity and uncertainty analysis algorithms.

See:
>> Description

Interface Summary

ComplexTable	A table that can hold any type of information.
ComplexTransformation	An interface for a transformation that is applied to a ComplexTable and produces a ComplexTable.
DoubleTable	This is the basic table for accessing floating point numbers.
DoubleTransformation	An interface for a transformation that is applied to a DoubleTable and produces a DoubleTable.
Executer	An Executer provides Operation execution queuing and control.
Operation	Operation represents a significant computation.
SelfDescribingOperation	An interface that allows Operations to describe themselves.
SimpleTable	This type of table can store floating point numbers (doubles), booleans, integers, and strings.
SimpleTransformation	An interface for a transformation that is applied to a SimpleTable and produces a SimpleTable.

Class Summary

ByReferenceBoolean	This provides a way for methods to return a boolean value through their argument list.
Column	This class describes a column in a data table.
RowException	Information about an exception related to a row in a table.

Exception Summary

TableException	An exception thrown when one of the semantics of operations on tables has been violated (e.g., close a table that has not been opened, access a table that has not been opened, access a value that has not been set).

2.1 Description

An overview and description of the COSU-API package may be found within the HTML-based electronic documentation, and is captured in the following format.

Calibration, Optimization, and Sensitivity and Uncertainty Analysis Algorithms Application Programming Interface (API)

Table of Contents

2.2 Goal

A number of groups develop tools or modeling frameworks that incorporate algorithms that drive repetitive execution of models. Such algorithms are used for purposes including sensitivity and uncertainty analysis, calibration of models, and optimization of parameters to best achieve one or more targets. The goal of this API is to allow the mathematical algorithms for sensitivity analysis, calibration, etc. to be implemented once, but used in multiple modeling tools and frameworks, even though those tools and frameworks do not share I/O or execution management approaches. This API could also be used as a common way to describe these algorithms, even if they are not actually implemented using this API.

The design group's hope is that functionalities implemented using this API would be shared with other developers and supported by the implementer. Commercial entities are also encouraged to develop proprietary capabilities that are expressed with the API or that use capabilities expressed with the API. Naturally, developers may choose to name collections of functionality that they develop.

2.3 Design Philosophy

The group that developed the API tried to adhere to the following design philosophies:

- The fewer the classes the better.
- Allow new algorithms to be added to modeling frameworks with little or no framework programming.
- Provide optional support for advanced capabilities (e.g., executing multiple instances of models in parallel).
- Support multiple platforms (e.g., Windows, Linux).

2.4 Language Choice

Java was chosen to express the API for the following reasons:

- Object-oriented concepts support extensibility, encapsulation, and explanation.
- Java is used by several of the modeling frameworks represented by members of the design group.
- Java is easier for new people to read and use than C++.
- Java can be interfaced with C relatively easily and in a platform-independent manner.

While Java has been used for the API, implementations of algorithms may use any language that can be interfaced with Java. For instance, computationally intensive algorithms could be written in C with a Java wrapper that conforms to the API.

For algorithms or applications where Java is not appropriate, the API could be considered as a general design and a corresponding API could be generated in another language. Tools are under development that will take the Java API and generate a substantial part of APIs in other languages, such as FORTRAN.

There are some disadvantages to expressing the API as Java. The following table summarizes the disadvantages and the methods that have been or will be used to address the disadvantages.

Disadvantage	Approach to Address
Object-oriented concepts can be difficult to express in FORTRAN.	Adapter code could be written that would ease the connection between the two approaches. Also, FORTRAN programs are most likely to use only a subset of the API that operates on floating point numbers, which eliminates some of the problems.
Java does not support generic collections.	The primary collection class, a table, in the API is expressed as several classes, each of which supports different data types. Java 1.5 will support generic collections. A consideration for the future is whether to extend the API to take advantage of that feature.

2.5 Design

Algorithms and models are represented by "Operations." Information used as inputs to and outputs from algorithms and models are passed in tables of various types.

One additional area that should be added to the design is an interface for distributions. This should include a way to specify random seeds and a way to obtain values from the distribution. An extension of the interface should allow inverse computations for the distribution.

An example approach taken toward this functionality can be found in FRAMES 3MRA 1.0. Its Windows-based modeling environment utilizes a "stat.dll" for similar statistical sampling. An "mc.dll" (i.e., Monte Carlo) provides a multi-language interface for calls to the "stat.dll."

2.6 Adoption Plan

The plan for adopting this API is as follows:

1. Solicit feedback from collaborators.
2. Incorporate feedback and redistribute design.
3. Implement the API for some algorithms and frameworks as a proof of concept.
4. Refine the API in light of lessons learned during the proof of concept.
5. Present the API at the uncertainty conference in August 2003 and solicit wider feedback and participation.

2.7 Original Design Group

The original design group for the API illustrated below included contributions from the following people:

- Steve Fine (US EPA)
- Karl Castleton (PNL)
- Ned Banta (USGS)
- Mary Hill (USGS)
- Steve Markstrom (USGS)
- George Leavesley (USGS)
- Justin Babendreier (US EPA)

Version: June 22, 2003

Package Class **Tree Deprecated Index Help**

PREV PACKAGE   NEXT PACKAGE   FRAMES   NO FRAMES   All Classes All Classes

3.0 Overview of Hierarchy for Package

The following object-oriented class structure summarizes the hierarchical scheme of the Calibration, Optimization, and Sensitivity and Uncertainty Analysis Algorithm Application Programming Interface (COSU-API).

3.1 Class Hierarchy

- class java.lang.Object
    - class org.iscmem.cosu.**ByReferenceBoolean**
    - class org.iscmem.cosu.**Column**
    - class org.iscmem.cosu.**RowException**
    - class java.lang.Throwable (implements java.io.Serializable)
        - class java.lang.Exception
            - class org.iscmem.cosu.**TableException**

3.2 Interface Hierarchy

- interface org.iscmem.cosu.**ComplexTransformation**
- interface org.iscmem.cosu.**DoubleTable**
    - interface org.iscmem.cosu.**SimpleTable**
        - interface org.iscmem.cosu.**ComplexTable**
- interface org.iscmem.cosu.**DoubleTransformation**
- interface org.iscmem.cosu.**Executer**
- interface org.iscmem.cosu.**Operation**
    - interface org.iscmem.cosu.**SelfDescribingOperation**
- interface org.iscmem.cosu.**SimpleTransformation**

3.3 File System Directory Structure

Documentation for the object oriented class package was formulated in HTML, and utilized the following analogous directory structure.

Archive: COSU_API.zip, archive size 48 Kb, decompressed size: 257 Kb, total 28 files.

Filename	Size	Packed	Modified	Path
allclasses-frame.html	1,913	634	6/22/2003 6:43 PM	
allclasses-noframe.html	1,783	626	6/22/2003 6:43 PM	
ByReferenceBoolean.html	8,722	1,745	6/22/2003 6:43 PM	org\iscmem\cosu\
Column.html	10,094	1,989	6/22/2003 6:43 PM	org\iscmem\cosu\
ComplexTable.html	17,883	2,778	6/22/2003 6:43 PM	org\iscmem\cosu\
ComplexTransformation.html	7,965	1,662	6/22/2003 6:43 PM	org\iscmem\cosu\
constant-values.html	6,569	1,240	6/22/2003 6:43 PM	
deprecated-list.html	4,177	923	6/22/2003 6:43 PM	
DoubleTable.html	26,514	3,838	6/22/2003 6:43 PM	org\iscmem\cosu\
DoubleTransformation.html	7,934	1,662	6/22/2003 6:43 PM	org\iscmem\cosu\
Executer.html	16,691	2,688	6/22/2003 6:43 PM	org\iscmem\cosu\
help-doc.html	7,411	2,179	6/22/2003 6:43 PM	
index-all.html	28,064	4,117	6/22/2003 6:43 PM	
index.html	716	436	6/22/2003 6:43 PM	
Operation.html	15,967	3,102	6/22/2003 6:43 PM	org\iscmem\cosu\
overview-tree.html	6,084	1,246	6/22/2003 6:43 PM	
package-frame.html	2,274	719	6/22/2003 6:43 PM	org\iscmem\cosu\
package-list	17	17	6/22/2003 6:43 PM	
package-summary.html	14,532	4,016	6/22/2003 6:43 PM	org\iscmem\cosu\
package-tree.html	6,328	1,234	6/22/2003 6:43 PM	org\iscmem\cosu\
packages.html	687	378	6/22/2003 6:43 PM	
RowException.html	7,516	1,594	6/22/2003 6:43 PM	org\iscmem\cosu\
SelfDescribingOperation.html	11,673	2,078	6/22/2003 6:43 PM	org\iscmem\cosu\
serialized-form.html	4,843	1,057	6/22/2003 6:43 PM	
SimpleTable.html	28,876	3,139	6/22/2003 6:43 PM	org\iscmem\cosu\
SimpleTransformation.html	7,814	1,654	6/22/2003 6:43 PM	org\iscmem\cosu\
stylesheet.css	1,328	442	6/22/2003 6:43 PM	
TableException.html	9,290	1,964	6/22/2003 6:43 PM	org\iscmem\cosu\

4.0 Detailed Hierarchy for Package

Details of interface object class structures are defined first, followed by a similar presentation for a group of support classes. The HTML-based documentation for all classes is generally represented by the following graphical menu structure.

Package `Class` **Tree Deprecated Index Help**
PREV CLASS **NEXT CLASS**   **FRAMES   NO FRAMES**   All Classes
SUMMARY: NESTED | FIELD | CONSTR | METHOD   DETAIL: FIELD | CONSTR | METHOD

4.1 Interface Classes

4.1.1 Interface DoubleTable

All Known Subinterfaces:
> ComplexTable, SimpleTable

---

public interface **DoubleTable**

This is the basic table for accessing floating point numbers. The table must be opened before it can be used. This allows tables to be implemented as files or via a database. Since the underlying mechanism for storing the contents of the table is not guaranteed to work (e.g., network connection lost), all methods that access the contents throw Exception.

At a minimum, information written to the table is assumed to be permanently stored in two situations:

1. When the table is closed (all calls to open have been matched with a call to close)
2. When values are written to a different row in the table. In other words, storing multiple values in the same row does not ensure that values in that row are permanently stored.

Implementations may choose to permanently store values in additional situations.

The table will maintain information about availability of values. If no value has been written to a cell, a value is not available. Attempting to retrieve such a value will cause a TableException to be thrown. Use *isDataAvailable* to determine if a value is present in a cell. The number of rows in a table will reflect the largest row number to which values have been stored.

Typically, table constructors will set the number of columns and additional information that determines where the contents of the table are stored (e.g., a file name). Implementations must ensure that all operations are thread-safe.

Method Summary

void	**close**() Indicate that access to a table is no longer required.
int	**findColumnByName**(java.lang.String colName) Return the 0-based index of the column with the given name or a negative number if the name was not found.
int	**findRowByName**(java.lang.String rowName) Return the 0-based index of the row with the given name or a negative number if the name was not found.
int	**getColumnCount**() Return the number of columns in the table
java.lang.String	**getColumnName**(int columnIndex) Return the name of a column
double	**getDoubleAt**(int rowIndex, int columnIndex) Return the number at the given row and column.
double[]	**getDoubles**(int rowIndex) Return the numbers in the given row, one for each column in the table.
int	**getRowCount**() Return the number of rows in the table
java.lang.String	**getRowName**(int rowIndex) Return the name for the given row, which might be null if the name has not been set.
boolean	**isCellEditable**(int rowIndex, int columnIndex) Indicate if the given cell in the table can be changed.
boolean	**isValueAvailable**(int rowIndex, int columnIndex) Indicate if a value has been stored in the given cell.
void	**open**() Indicate that the contents of the table will be accessed.
void	**setDoubleAt**(int rowIndex, int columnIndex, double aValue) Put a value in the table.
void	**setDoubles**(int rowIndex, double[] values) Fill a row in the table.
void	**setRowName**(int rowIndex, java.lang.String name) Set the name for the given row.
void	**waitForDataAvailable**(int rowIndex, int columnIndex) Wait for a value to become available in the given cell.

Method Detail

open

public void **open**()
        throws java.lang.Exception

Indicate that the contents of the table will be accessed. This must be called before any operation that access the contents. Each call to "open" must be matched to a call to "close". It is permissible to call "open" while the table is already open.

Throws:
java.lang.Exception - if there are problems accessing the table's contents

---

close

public void **close**()
        throws java.lang.Exception

Indicate that access to a table is no longer required. Each call to "close" balances a call to "open". If the table is not open, a TableException will be thrown.

Throws:
java.lang.Exception - if there are problems accessing the table's contents

---

getColumnCount

public int **getColumnCount**()
        throws java.lang.Exception

Return the number of columns in the table.

Throws:
TableException - if the table is not open
java.lang.Exception - if there are problems accessing the table's contents

---

getColumnName

public java.lang.String **getColumnName**(int columnIndex)
        throws java.lang.Exception

Return the name of a column.

Parameters:
columnIndex - 0-based column index

Throws:
TableException - if the table is not open
java.lang.Exception - if there are problems accessing the table's contents
java.lang.IllegalArgumentException - if the index is out of range

getRowCount

public int **getRowCount**()
                throws java.lang.Exception

Return the number of rows in the table.

Throws:
TableException - if the table is not open
java.lang.Exception - if there are problems accessing the table's contents

isValueAvailable

public boolean **isValueAvailable**(int rowIndex,
                                int columnIndex)
                throws java.lang.Exception

Indicate if a value has been stored in the given cell.

Parameters:
rowIndex - 0-based row index
columnIndex - 0-based column index

Throws:
java.lang.IllegalArgumentException - if an index is out of range
TableException - if the table is not open
java.lang.Exception - if there are problems accessing the table's contents

waitForDataAvailable

public void **waitForDataAvailable**(int rowIndex,
                                int columnIndex)
                throws java.lang.Exception

Wait for a value to become available in the given cell.

Parameters:
rowIndex - 0-based row index
columnIndex - 0-based column index

Throws:
java.lang.IllegalArgumentException - if an index is out of range
java.lang.InterruptedException - if the thread is interrupted while waiting

TableException - if the table is not open
java.lang.Exception - if there are problems accessing the table's contents

getDoubleAt

public double **getDoubleAt**(int rowIndex,
                    int columnIndex)
            throws java.lang.Exception

Return the number at the given row and column. The number might be Double.NaN.

Parameters:
rowIndex - 0-based row index
columnIndex - 0-based column index

Throws:
java.lang.IllegalArgumentException - if an index is out of range
TableException - if the table is not open or if the column does not contain a double (for derived table classes)
java.lang.Exception - if there are problems accessing the table's contents

getDoubles

public double[] **getDoubles**(int rowIndex)
                throws java.lang.Exception

Return the numbers in the given row, one for each column in the table. The numbers might be Double.NaN.

Parameters:
rowIndex - 0-based row index

Throws:
java.lang.IllegalArgumentException - if an index is out of range
TableException - if the table is not open or if all columns do not contain a double (for derived table classes)
java.lang.Exception - if there are problems accessing the table's contents

isCellEditable

public boolean **isCellEditable**(int rowIndex,
                        int columnIndex)
                throws java.lang.Exception

Indicate if the given cell in the table can be changed.

Parameters:
rowIndex - 0-based row index
columnIndex - 0-based column index

Throws:
java.lang.IllegalArgumentException - if an index is out of range
TableException - if the table is not open
java.lang.Exception - if there are problems accessing the table's contents

getRowName

public java.lang.String **getRowName**(int rowIndex)
                        throws java.lang.Exception

Return the name for the given row, which might be null if the name has not been set.

Parameters:
rowIndex - 0-based row index

Throws:
java.lang.IllegalArgumentException - if an index is out of range
java.lang.Exception - if there are problems accessing the table's contents

setRowName

public void **setRowName**(int rowIndex,
                        java.lang.String name)
                throws java.lang.Exception

Set the name for the given row.

Parameters:
rowIndex - 0-based row index
name - String the name of the row

Throws:
java.lang.IllegalArgumentException - if an index is out of range
TableException - if the table is not open
java.lang.Exception - if there are problems accessing the table's contents

findRowByName

public int **findRowByName**(java.lang.String rowName)
                throws java.lang.Exception

Return the 0-based index of the row with the given name or a negative number if the name was not found. Name comparisons will respect case.

Throws:
TableException - if the table is not open
java.lang.Exception - if there are problems accessing the table's contents

findColumnByName

public int **findColumnByName**(java.lang.String colName)
        throws java.lang.Exception

> Return the 0-based index of the column with the given name or a negative number if the name was not found. Name comparisons will respect case.
>
> **Throws:**
> TableException - if the table is not open
> java.lang.Exception - if there are problems accessing the table's contents

---

setDoubleAt

public void **setDoubleAt**(int rowIndex,
                int columnIndex,
                double aValue)
        throws java.lang.Exception

> Put a value in the table. Any values previously stored in other rows will be written to persistent storage.
>
> **Parameters:**
> rowIndex - 0-based row index
> columnIndex - 0-based column index
> aValue - double value for the cell
>
> **Throws:**
> java.lang.IllegalArgumentException - if an index is out of range
> TableException - if the table is not open
> java.lang.Exception - if there are problems accessing the table's contents

---

setDoubles

public void **setDoubles**(int rowIndex,
                double[] values)

> Fill a row in the table. Any values previously stored in other rows will be written to persistent storage.
>
> **Parameters:**
> rowIndex - 0-based row index
> values - double[] containing a value for each column
>
> **Throws:**
> java.lang.IllegalArgumentException - if an index is out of range
> TableException - if the table is not open
> java.lang.Exception - if there are problems accessing the table's contents

4.1.2 Interface SimpleTable

All Superinterfaces:
   DoubleTable

All Known Subinterfaces:
   ComplexTable

---

public interface **SimpleTable**
extends DoubleTable

This type of table can store floating point numbers (doubles), booleans, integers, and strings.

Field Summary

static int	**BOOLEAN**
static int	**DOUBLE**
static int	**INTEGER**
static int	**OBJECT**
static int	**STRING**

Method Summary

boolean	**getBooleanAt**(int rowIndex, int columnIndex) Return the value in the given cell.
boolean[]	**getBooleans**(int rowIndex) Return the values in the given row.
int	**getColumnType**(int columnIndex) Return a code indicating the type of information stored in the given column.
int	**getIntAt**(int rowIndex, int columnIndex) Return the value in the given cell.
int[]	**getInts**(int rowIndex) Return the values in the given row.
java.lang.String	**getStringAt**(int rowIndex, int columnIndex) Return the value in the given cell.
java.lang.String[]	**getStrings**(int rowIndex) Return the values in the given row.
void	**setBooleanAt**(int rowIndex, int columnIndex, boolean aValue) Set the value in the given cell.
void	**setBooleans**(int rowIndex, boolean[] values) Set the values in the given row.
void	**setIntAt**(int rowIndex, int columnIndex, int aValue) Set the value in the given cell.

void	setInts(int rowIndex, int[] values)	
	Set the values in the given row.	
void	setStringAt(int rowIndex, int columnIndex, java lang.String aValue)	
	Set the value in the given cell.	
void	setStrings(int rowIndex, java lang.String[] values)	
	Set the values in the given row.	

Methods inherited from interface org.iscmem.cosu.DoubleTable

close, findColumnByName, findRowByName, getColumnCount, getColumnName, getDoubleAt, getDoubles, getRowCount, getRowName, isCellEditable, isValueAvailable, open, setDoubleAt, setDoubles, setRowName, waitForDataAvailable

Field Detail

BOOLEAN

public static final int **BOOLEAN**
See Also:
> Constant Field Values

INTEGER

public static final int **INTEGER**

See Also:
> Constant Field Values

DOUBLE

public static final int **DOUBLE**

See Also:
> Constant Field Values

STRING

public static final int **STRING**

See Also:
> Constant Field Values

OBJECT

public static final int **OBJECT**

See Also:
    Constant Field Values

Method Detail

getColumnType

public int **getColumnType**(int columnIndex)
        throws java.lang.Exception

Return a code indicating the type of information stored in the given column.

Parameters:
columnIndex - 0-based index of the column

Throws:
java.lang.IllegalArgumentException - if the index is out of range
java.lang.Exception - if there are problems accessing the table's contents

---

getIntAt

public int **getIntAt**(int rowIndex,
          int columnIndex)
        throws java.lang.Exception

Return the value in the given cell.

Parameters:
rowIndex - 0-based row index
columnIndex - 0-based column index

Throws:
java.lang.IllegalArgumentException - if an index is out of range
TableException - if the table is not open or if the column does not contain the expected type
java.lang.Exception - if there are problems accessing the table's contents

---

getBooleanAt

public boolean **getBooleanAt**(int rowIndex,
                           int columnIndex)
                throws java.lang.Exception

Return the value in the given cell.

Parameters:
rowIndex - 0-based row index
columnIndex - 0-based column index

Throws:
java.lang.IllegalArgumentException - if an index is out of range
TableException - if the table is not open or if the column does not contain the expected type
java.lang.Exception - if there are problems accessing the table's contents

---

getStringAt

public java.lang.String **getStringAt**(int rowIndex,
                           int columnIndex)
                throws java.lang.Exception

Return the value in the given cell.

Parameters:
rowIndex - 0-based row index
columnIndex - 0-based column index

Throws:
java.lang.IllegalArgumentException - if an index is out of range
TableException - if the table is not open or if the column does not contain the expected type
java.lang.Exception - if there are problems accessing the table's contents

---

setIntAt

public void **setIntAt**(int rowIndex,
                int columnIndex,
                int aValue)
            throws java.lang.Exception

Set the value in the given cell.

Parameters:
rowIndex - 0-based row index
columnIndex - 0-based column index
aValue - the int to store

Throws:
java.lang.IllegalArgumentException - if an index is out of range
TableException - if the table is not open or if the column does not contain the expected type
java.lang.Exception - if there are problems accessing the table's contents

---

setBooleanAt

public void **setBooleanAt**(int rowIndex,
                                 int columnIndex,
                                 boolean aValue)
                         throws java.lang.Exception

Set the value in the given cell.

Parameters:
rowIndex - 0-based row index
columnIndex - 0-based column index
aValue - the boolean to store

Throws:
java.lang.IllegalArgumentException - if an index is out of range
TableException - if the table is not open or if the column does not contain the expected type
java.lang.Exception - if there are problems accessing the table's contents

---

setStringAt

public void **setStringAt**(int rowIndex,
                              int columnIndex,
                              java.lang.String aValue)
                      throws java.lang.Exception

Set the value in the given cell.

Parameters:
rowIndex - 0-based row index
columnIndex - 0-based column index
aValue - the String to store

Throws:
java.lang.IllegalArgumentException - if an index is out of range
TableException - if the table is not open or if the column does not contain the expected type
java.lang.Exception - if there are problems accessing the table's contents

---

getInts

public int[] **getInts**(int rowIndex)
                    throws java.lang.Exception

    Return the values in the given row.

    Parameters:
    rowIndex - 0-based row index

    Throws:
    java.lang.IllegalArgumentException - if an index is out of range
    TableException - if the table is not open or if all of the columns do not contain the expected type
    java.lang.Exception - if there are problems accessing the table's contents

---

getBooleans

public boolean[] **getBooleans**(int rowIndex)
                        throws java.lang.Exception

    Return the values in the given row.

    Parameters:
    rowIndex - 0-based row index

    Throws:
    java.lang.IllegalArgumentException - if an index is out of range
    TableException - if the table is not open or if all of the columns do not contain the expected type
    java.lang.Exception - if there are problems accessing the table's contents

---

getStrings

public java.lang.String[] **getStrings**(int rowIndex)
                            throws java.lang.Exception

    Return the values in the given row.

    Parameters:
    rowIndex - 0-based row index

    Throws:
    java.lang.IllegalArgumentException - if an index is out of range
    TableException - if the table is not open or if all of the columns do not contain the expected type
    java.lang.Exception - if there are problems accessing the table's contents

---

setInts

public void **setInts**(int rowIndex,
                            int[] values)
                    throws java.lang.Exception

Set the values in the given row.

Parameters:
rowIndex - 0-based row index
values - the ints for the entire row

Throws:
java.lang.IllegalArgumentException - if an index is out of range
TableException - if the table is not open, if all of the columns do not contain the expected type, or if
values.length != # of columns
java.lang.Exception - if there are problems accessing the table's contents

---

setBooleans

public void **setBooleans**(int rowIndex,
                            boolean[] values)
                    throws java.lang.Exception

Set the values in the given row.

Parameters:
rowIndex - 0-based row index
values - the booleans for the entire row

Throws:
java.lang.IllegalArgumentException - if an index is out of range
TableException - if the table is not open, if all of the columns do not contain the expected type, or if
                            values.length != # of columns
java.lang.Exception - if there are problems accessing the table's contents

---

setStrings

public void **setStrings**(int rowIndex,
                            java.lang.String[] values)
                    throws java.lang.Exception

Set the values in the given row.

Parameters:
rowIndex - 0-based row index
values - the strings for the entire row

Throws:
java.lang.IllegalArgumentException - if an index is out of range
TableException - if the table is not open, if all of the columns do not contain the expected type, or if values.length != # of columns
java.lang.Exception - if there are problems accessing the table's content

4.1.3 Interface ComplexTable

All Superinterfaces:
    DoubleTable, SimpleTable

---

public interface **ComplexTable**
extends SimpleTable

A table that can hold any type of information. This extends tables that hold primitive types with the ability to hold Objects.

Field Summary

Fields inherited from interface org.iscmem.cosu.SimpleTable

BOOLEAN, DOUBLE, INTEGER, OBJECT, STRING

Method Summary

Column[]	**getColumns**() Return information about all of the column in the table.
java.lang.Object	**getObjectAt**(int rowIndex, int columnIndex) Return the value in the given cell.
java.lang.Object[]	**getObjects**(int rowIndex) Return the values in the given row.
void	**setObjectAt**(int rowIndex, int columnIndex, java.lang.Object aValue) Set the value in the given cell.
void	**setObjects**(int rowIndex, java.lang.Object[] values) Set the values in the given row.

Methods inherited from interface org.iscmem.cosu.SimpleTable

getBooleanAt, getBooleans, getColumnType, getIntAt, getInts, getStringAt, getStrings, setBooleanAt, setBooleans, setIntAt, setInts, setStringAt, setStrings

Method Detail

getColumns

public <u>Column</u>[] **getColumns**()
             throws java.lang.Exception

Return information about all of the column in the table. The information includes the column names and metadata about each column. At this time, the meta-data will be the Class that the column holds, except int, double, and boolean will be represented by Int, Double, and Boolean.

Throws:
<u>TableException</u> - if the table is not open
java.lang.Exception - if there are problems accessing the table's contents

---

getObjectAt

public java.lang.Object **getObjectAt**(int rowIndex,
                           int columnIndex)
             throws java.lang.Exception

Return the value in the given cell.

Parameters:
rowIndex - 0-based row index
columnIndex - 0-based column index

Throws:
java.lang.IllegalArgumentException - if an index is out of range
<u>TableException</u> - if the table is not open or if the column does not contain the expected type
java.lang.Exception - if there are problems accessing the table's contents

---

setObjectAt

public void **setObjectAt**(int rowIndex,
               int columnIndex,
               java.lang.Object aValue)
         throws java.lang.Exception

Set the value in the given cell.

Parameters:
rowIndex - 0-based row index
columnIndex - 0-based column index
aValue - the Object to store

Throws:
java.lang.IllegalArgumentException - if an index is out of range
TableException - if the table is not open or if the column does not contain the expected type
java.lang.Exception - if there are problems accessing the table's contents

---

getObjects

public java.lang.Object[] **getObjects**(int rowIndex)
                               throws java.lang.Exception

Return the values in the given row.

Parameters:
rowIndex - 0-based row index

Throws:
java.lang.IllegalArgumentException - if an index is out of range
TableException - if the table is not open or if all of the columns do not contain the expected type
java.lang.Exception - if there are problems accessing the table's contents

---

setObjects

public void **setObjects**(int rowIndex,
                       java.lang.Object[] values)
                 throws java.lang.Exception

Set the values in the given row.

Parameters:
rowIndex - 0-based row index
values - the Objects for the entire row

Throws:
java.lang.IllegalArgumentException - if an index is out of range
TableException - if the table is not open, if all of the columns do not contain the expected type, or if
                     values.length != # of columns
java.lang.Exception - if there are problems accessing the table's contents

4.1.4 Interface DoubleTransformation

public interface **DoubleTransformation**

An interface for a transformation that is applied to a DoubleTable and produces a DoubleTable. The sizes of the input and output tables need not be the same. Examples of transformations include extracting subsets and aggregating information.

Method Summary

DoubleTable	**transform**(DoubleTable input)
	Apply the transformation and return the result.

Method Detail

transform

public <u>DoubleTable</u> **transform**(<u>DoubleTable</u> input)
                throws java.lang.Exception

Apply the transformation and return the result.

Parameters:
input - DoubleTable that is input for the transformation

4.1.5 Interface SimpleTransformation

public interface **SimpleTransformation**

An interface for a transformation that is applied to a SimpleTable and produces a SimpleTable. The sizes of the input and output tables need not be the same. Examples of transformations include extracting subsets and aggregating information.

Method Summary

SimpleTable	**transform**(SimpleTable input)
	Apply the transformation and return the result.

Method Detail

transform

public <u>SimpleTable</u> **transform**(<u>SimpleTable</u> input)
　　　　　　　　throws java.lang.Exception

> Apply the transformation and return the result.
>
> **Parameters:**
> input - SimpleTable that is input for the transformation

4.1.6 Interface ComplexTransformation

public interface **ComplexTransformation**

An interface for a transformation that is applied to a ComplexTable and produces a ComplexTable. The sizes of the input and output tables need not be the same. Examples of transformations include extracting subsets and aggregating information.

Method Summary

ComplexTable	**transform**(ComplexTable input) 　　　　　Apply the transformation and return the result.

Method Detail

transform

public <u>ComplexTable</u> **transform**(<u>ComplexTable</u> input)
　　　　　throws java.lang.Exception

> Apply the transformation and return the result.
>
> **Parameters:**
> input - ComplexTable that is input for the transformation

4.1.7 Interface Operation

All Known Subinterfaces:
SelfDescribingOperation

---

public interface **Operation**

Operation represents a significant computation. Examples include models and calibration, optimization, and sensitivity and uncertainty analysis algorithms. Operations are typically more complex and substantial computations than Transformations. An Operation provides several additional capabilities over Transformations:

- Support for restarting long computations that were interrupted
- Set up before and clean up after a number of executions of an Operation
- Indication of whether Operations can be executed in parallel
- A method to prematurely terminate computations

The typical way to use an Operation is to call *setUp*, call run one or more times, and finally call *cleanUp*.

Run and *restart* return DoubleTable, which is the superclass of the other table types. If the calling routine is expecting a more complex type of table to be returned, it can check the actual type of the returned table.

Operations, such as a Monte Carlo algorithm, that will invoke other Operations may accept an Executer as an argument to take advantage of parallel execution facilities that some modeling systems might provide.

Method Summary

boolean	**canRestart**() Indicate if the Operation can be restarted with partial results from a previous invocation.
void	**cleanUp**() After the final call to run or restart, cleanUp must be called to provide an opportunity to perform any finally housecleaning that is required.
DoubleTable	**restart**(DoubleTable input, DoubleTable partialResult, ByReferenceBoolean complete) Restart the operation using the partial results returned from a previous invocation.
DoubleTable	**run**(DoubleTable input, ByReferenceBoolean complete) Perform the Operation.
void	**setUp**() Before calling run or restart the first time, setUp should be called.
void	**stop**() Request that the Operation stop prematurely.
boolean	**supportsParallelRuns**() Indicate if multiple instances of the Operation can be performed in parallel.

Method Detail

setUp

public void **setUp**()
    throws java.lang.Exception

> Before calling *run* or *restart* the first time, *setUp* should be called. This provides an opportunity to perform any one-time configuration before one or more executions of the operation.

cleanUp

public void **cleanUp**()
    throws java.lang.Exception

> After the final call to *run* or *restart*, *cleanUp* must be called to provide an opportunity to perform any final housecleaning that is required.

run

public DoubleTable **run**(DoubleTable input,
                    ByReferenceBoolean complete)
    throws java.lang.Exception

> Perform the operation.
>
> ### Parameters:
> input - DoubleTable containing the inputs for the operation. The table might be derived from DoubleTable.
>   Each operation implementation should confirm that the proper type of table has been provided.
> complete - ByReferenceBoolean that on return will indicate whether the operation was completed.
>
> ### Returns:
> A table containing the results. If the operation was prematurely terminated, this might be the partial results that were completed.

canRestart

public boolean **canRestart**()

> Indicate if the Operation can be restarted with partial results from a previous invocation.

restart

public <u>DoubleTable</u> **restart**(<u>DoubleTable</u> input,
        <u>DoubleTable</u> partialResult,
        <u>ByReferenceBoolean</u> complete)
     throws java.lang.Exception

Restart the Operation using the partial results returned from a previous invocation.

Parameters:
input - DoubleTable containing the original inputs for the operation. The table might be derived from
  DoubleTable. Each operation implementation should confirm that the proper type of table has been
  provided.
complete - ByReferenceBoolean that on return will indicate whether the operation was completed.

Returns:
A table containing the results. If the Operation was prematurely terminated, this might be the partial results
that were completed.

---

supportsParallelRuns

public boolean **supportsParallelRuns**()

Indicate if multiple instances of the Operation can be performed in parallel. For instance, if the Operation represents an external program that produces output files in a fixed location, then the result would be *false* since the outputs might overwrite each other.

---

stop

public void **stop**()
   throws java.lang.Exception

Request that the Operation stop prematurely. Operations need not provide this service (in which case, the implementation is an empty method body), so the caller should not assume that computations will cease immediately.

4.1.8 Interface SelfDescribingOperation

All Superinterfaces:
   Operation

---

public interface **SelfDescribingOperation**
extends Operation

An interface that allows Operations to describe themselves. Implementations of an Operation can choose to implement this interface to allow frameworks to present the user information about the Operation and to request information from the user.

Method Summary

java.lang.String	**getDescription**() Return a description of the Operation.
Column[]	**getInputColumns**() Return information about the columns the Operation expects to see as input.
java.lang.String	**getName**() Return a human-meaningful name of the Operation.
Column[]	**getOutputColumns**() Return information about the columns the Operation will produce.
java net.URL	**getURL**() Return the home page for the Operation or null if none.

Methods inherited from interface org.iscmem.cosu.Operation

canRestart, cleanUp, restart, run, setUp, stop, supportsParallelRuns

---

Method Detail

getDescription

public java.lang.String **getDescription**()

   Return a description of the Operation.

---

getName

public java.lang.String **getName**()

   Return a human-meaningful name of the Operation.

---

getURL

public java.net.URL **getURL**()

> Return the home page for the Operation or null if none.

---

getInputColumns

public Column[] **getInputColumns**()

> Return information about the columns the Operation expects to see as input.

---

getOutputColumns

public Column[] **getOutputColumns**()

> Return information about the columns the Operation will produce.

4.1.9 Interface Executer

public interface **Executer**

An Executer provides Operation execution queuing and control. Modeling systems provide one or more implementations of the Executer that control how execution is performed. A simple Executer executes everything in serial. More complex Executers may execute Operations in parallel to take advantage of multiple CPUs.

Each operation that is queued will be executed. Then, if a transformation was supplied, it will be executed. Finally, the first row of the resulting table will be added to an aggregate result table that is being accumulated.

The purpose of including an Executer in the design is to allow iterative operations, such as a Monte Carlo algorithm, to be written in a manner that can take advantage of task parallelism without requiring each algorithm writer to implement their own multithreaded execution management system.

Implementations must be thread-safe.

Method Summary

RowException	**getException**(int index) Return Exception information
int	**getExceptionCount**() Return how many Exceptions have been thrown by operations.
DoubleTable	**getResult**() Return the aggregate table where results have been accumulated.
boolean	**isDone**() Indicate if all queued Operations have been completed.
void	**queue**(Operation op, int rowIndex) Queue an Operation to be run with no Transformation.
void	**queue**(Operation op, int rowIndex, ComplexTransformation xform) Queue an Operation to be run with an optional Transformation.
void	**queue**(Operation op, int rowIndex, DoubleTransformation xform) Queue an Operation to be run with an optional Transformation.
void	**queue**(Operation op, int rowIndex, SimpleTransformation xform) Queue an Operation to be run with an optional Transformation.
void	**stopExecution**() Stop execution to the extent feasible.
void	**waitForDone**() Return when all queued Operations have completed.

Method Detail

queue

public void **queue**(Operation op,
            int rowIndex)

Queue an Operation to be run with no Transformation.

Parameters:
op - Operation to be executed
rowIndex - 0-based index of the row where the result should appear in the aggregate table

---

queue

public void **queue**(Operation op,
            int rowIndex,
            DoubleTransformation xform)

Queue an Operation to be run with an optional Transformation.

Parameters:
op - Operation to be executed
rowIndex - 0-based index of the row where the result should appear in the aggregate table
xform - DoubleTransformation to be applied to result of Operation

queue

public void **queue**(Operation op,
　　　　　　int rowIndex,
　　　　　　SimpleTransformation xform)

Queue an Operation to be run with an optional Transformation.

Parameters:
op - Operation to be executed
rowIndex - 0-based index of the row where the result should appear in the aggregate table
xform - SimpleTransformation to be applied to result of Operation

queue

public void **queue**(Operation op,
　　　　　　int rowIndex,
　　　　　　ComplexTransformation xform)

Queue an Operation to be run with an optional Transformation.

Parameters:
op - Operation to be executed
rowIndex - 0-based index of the row where the result should appear in the aggregate table
xform - ComplexTransformation to be applied to result of Operation

isDone

public boolean **isDone**()

Indicate if all queued Operations have been completed.

waitForDone

public void **waitForDone**()

Return when all queued Operations have completed.

getResult

public <u>DoubleTable</u> **getResult**()

> Return the aggregate table where results have been accumulated.

---

stopExecution

public void **stopExecution**()
> throws java.lang.Exception

> Stop execution to the extent feasible. Some implementations may choose to do nothing.

---

getExceptionCount

public int **getExceptionCount**()

> Return how many Exceptions have been thrown by Operations.

---

getException

public <u>RowException</u> **getException**(int index)

> Return Exception information

> **Parameters:**
> index - the 0-based number of the Exception to return

4.2 Support Classes

4.2.1 Class ByReferenceBoolean

java.lang.Object

  |
  +--**org.iscmem.cosu.ByReferenceBoolean**

---

public class **ByReferenceBoolean**
extends java.lang.Object

This provides a way for methods to return a boolean value through their argument list. It is useful when a method must return two values.

Field Summary

Boolean	value

Constructor Summary

ByReferenceBoolean()

Methods inherited from class java.lang.Object

clone, equals, finalize, getClass, hashCode, notify, notifyAll, toString, wait, wait, wait

Field Detail

value

public boolean **value**

Constructor Detail

ByReferenceBoolean

public **ByReferenceBoolean**()

4.2.2 Class Column

java.lang.Object
  |
  +--org iscmem.cosu.Column

---

public class **Column**
extends java.lang.Object

This class describes a column in a data table.

Field Summary

java.lang.String	**description** An optional description of the column or null if not provided.
java.lang.String	**name** The name of the column.
java.lang.Object	**type** Information about the type of information stored in the column.

Constructor Summary

Column()

Methods inherited from class java.lang.Object

clone, equals, finalize, getClass, hashCode, notify, notifyAll, toString, wait, wait, wait

Field Detail

name

public java.lang.String **name**

The name of the column.

---

type

public java.lang.Object **type**

Information about the type of information stored in the column. Initially, this will be of type Class. In the future, this might return an object that contains additional metadata. Int, Double, and Boolean will be used to represent columns containing int, double, and boolean, respectively.

description

public java.lang.String **description**

An optional description of the column or null if not provided.

Constructor Detail

Column

public **Column**()

4.2.3 Class RowException

java.lang.Object
```
 |
 +--org.iscmem.cosu.RowException
```

public class **RowException**
extends java.lang.Object

Information about an exception related to a row in a table.

Constructor Summary

RowException()

Methods inherited from class java.lang.Object

clone, equals, finalize, getClass, hashCode, notify, notifyAll, toString, wait, wait, wait

Constructor Detail

RowException

public **RowException**()

4.2.4 Class TableException

```
java.lang.Object
  |
  +--java.lang.Throwable
       |
       +--java.lang.Exception
            |
            +--org.iscmem.cosu.TableException
```

All Implemented Interfaces:
      java.io.Serializable

---

public class **TableException**
extends java.lang.Exception

An Exception thrown when one of the semantics of Operations on tables has been violated (e.g., close a table that has not been opened, access a table that has not been opened, access a value that has not been set).

See Also:
      Serialized Form

Constructor Summary

TableException()   Creates a new instance of TableException without detail message.
TableException(java.lang.String msg)   Constructs an instance of TableException with the specified detail message.

Methods inherited from class java.lang.Throwable

fillInStackTrace, getCause, getLocalizedMessage, getMessage, getStackTrace, initCause, printStackTrace, printStackTrace, printStackTrace, setStackTrace, toString

---

Methods inherited from class java.lang.Object

clone, equals, finalize, getClass, hashCode, notify, notifyAll, wait, wait, wait

---

Constructor Detail

TableException

public **TableException**()

    Creates a new instance of *TableException* without detail message.

TableException

public **TableException**(java.lang.String msg)

Constructs an instance of *TableException* with the specified detail message.

Parameters:
msg - the detail message.

4.2.4.1 Serialized Form

Package org.iscmem.cosu

Class org.iscmem.cosu.TableException implements Serializable

Package Class **Tree Deprecated Index Help**

PREV   NEXT   FRAMES   NO FRAMES   All Classes All Classes

5.0 Deprecated API

Currently there are no deprecations of the API.

Package Class **Tree** **Deprecated** **Index** **Help**

PREV NEXT   FRAMES NO FRAMES   All Classes

6.0 Index

B

BOOLEAN - Static variable in interface org.iscmem.cosu.SimpleTable

ByReferenceBoolean - class org.iscmem.cosu.ByReferenceBoolean.
This provides a way for methods to return a boolean value through their argument list.

ByReferenceBoolean() - Constructor for class org.iscmem.cosu.ByReferenceBoolean

---

C

canRestart() - Method in interface org.iscmem.cosu.Operation
Indicate if the Operation can be restarted with partial results from a previous invocation.

cleanUp() - Method in interface org.iscmem.cosu.Operation
After the final call to run or restart, cleanUp must be called to provide an opportunity to perform any finally housecleaning that is required.

close() - Method in interface org.iscmem.cosu.DoubleTable
Indicate that access to a table is no longer required.

Column - class org.iscmem.cosu.Column.
This class describes a column in a data table.

Column() - Constructor for class org.iscmem.cosu.Column

ComplexTable - interface org.iscmem.cosu.ComplexTable.
A table that can hold any type of information.

ComplexTransformation - interface org.iscmem.cosu.ComplexTransformation.
An interface for a transformation that is applied to a ComplexTable and produces a ComplexTable.

D

description - Variable in class org.iscmem.cosu.Column
> An optional description of the column or null if not provided.

DOUBLE - Static variable in interface org.iscmem.cosu.SimpleTable

DoubleTable - interface org.iscmem.cosu.DoubleTable.
> This is the basic table for accessing floating point numbers.

DoubleTransformation - interface org.iscmem.cosu.DoubleTransformation.
> An interface for a Transformation that is applied to a DoubleTable and produces a DoubleTable.

E

Executer - interface org.iscmem.cosu.Executer.
> An Executor provides Operation execution queuing and control.

F

findColumnByName(String) - Method in interface org.iscmem.cosu.DoubleTable
> Return the 0-based index of the column with the given name or a negative number if the name was not found.

findRowByName(String) - Method in interface org.iscmem.cosu.DoubleTable
> Return the 0-based index of the row with the given name or a negative number if the name was not found.

G

getBooleanAt(int, int) - Method in interface org.iscmem.cosu.SimpleTable
> Return the value in the given cell.

getBooleans(int) - Method in interface org.iscmem.cosu.SimpleTable
> Return the values in the given row.

getColumnCount() - Method in interface org.iscmem.cosu.DoubleTable
> return the number of columns in the table.

getColumnName(int) - Method in interface org.iscmem.cosu.DoubleTable
    Return the name of a column.

getColumns() - Method in interface org.iscmem.cosu.ComplexTable
    Return information about all of the column in the table.

getColumnType(int) - Method in interface org.iscmem.cosu.SimpleTable
    Return a code indicating the type of information stored in the given column.

getDescription() - Method in interface org.iscmem.cosu.SelfDescribingOperation
    Return a description of the Operation.

getDoubleAt(int, int) - Method in interface org.iscmem.cosu.DoubleTable
    Return the number at the given row and column.

getDoubles(int) - Method in interface org.iscmem.cosu.DoubleTable
    Return the numbers in the given row, one for each column in the table.

getException(int) - Method in interface org.iscmem.cosu.Executer
    Return Exception information.

getExceptionCount() - Method in interface org.iscmem.cosu.Executer
    Return how many Exceptions have been thrown by Operations.

getInputColumns() - Method in interface org.iscmem.cosu.SelfDescribingOperation
    Return information about the columns the Operation expects to see as input.

getIntAt(int, int) - Method in interface org.iscmem.cosu.SimpleTable
    Return the value in the given cell.

getInts(int) - Method in interface org.iscmem.cosu.SimpleTable
    Return the values in the given row.

getName() - Method in interface org.iscmem.cosu.SelfDescribingOperation
    Return a human-meaningful name of the Operation.

getObjectAt(int, int) - Method in interface org.iscmem.cosu.ComplexTable
    Return the value in the given cell.

getObjects(int) - Method in interface org.iscmem.cosu.ComplexTable
    Return the values in the given row.

getOutputColumns() - Method in interface org.iscmem.cosu.SelfDescribingOperation
    Return information about the columns the Operation will produce.

getResult() - Method in interface org.iscmem.cosu.Executer
    Return the aggregate table where results have been accumulated.

getRowCount() - Method in interface org.iscmem.cosu.DoubleTable
    Return the number of rows in the table.

getRowName(int) - Method in interface org.iscmem.cosu.DoubleTable
    Return the name for the given row, which might be null if the name has not been set.

getStringAt(int, int) - Method in interface org.iscmem.cosu.SimpleTable
    Return the value in the given cell.

getStrings(int) - Method in interface org.iscmem.cosu.SimpleTable
    Return the values in the given row.

getURL() - Method in interface org.iscmem.cosu.SelfDescribingOperation
    Return the home page for the Operation or null if none.

I

INTEGER - Static variable in interface org.iscmem.cosu.SimpleTable

isCellEditable(int, int) - Method in interface org.iscmem.cosu.DoubleTable
    Indicate if the given cell in the table can be changed.

isDone() - Method in interface org.iscmem.cosu.Executer
    Indicate if all queued Operations have been completed.

isValueAvailable(int, int) - Method in interface org.iscmem.cosu.DoubleTable
    Indicate if a value has been stored in the given cell.

N

name - Variable in class org.iscmem.cosu.Column
    The name of the column.

O

OBJECT - Static variable in interface org.iscmem.cosu.SimpleTable

open() - Method in interface org.iscmem.cosu.DoubleTable
    Indicate that the contents of the table will be accessed.

Operation - interface org.iscmem.cosu.Operation.
Operation represents a significant computation.

org.iscmem.cosu - package org.iscmem.cosu
An Application Programming Interface (API) for calibration, optimization, and
sensitivity and uncertainty analysis algorithms.

Q

queue(Operation, int) - Method in interface org.iscmem.cosu.Executer
Queue an Operation to be run with no Transformation.

queue(Operation, int, ComplexTransformation) - Method in interface
org.iscmem.cosu.Executer
Queue an Operation to be run with an optional Transformation.

queue(Operation, int, DoubleTransformation) - Method in interface
org.iscmem.cosu.Executer
Queue an Operation to be run with an optional Transformation.

queue(Operation, int, SimpleTransformation) - Method in interface
org.iscmem.cosu.Executer
Queue an Operation to be run with an optional Transformation.

R

restart(DoubleTable, DoubleTable, ByReferenceBoolean) - Method in interface
org.iscmem.cosu.Operation
Restart the Operation using the partial results returned from a previous invocation.

RowException - class org.iscmem.cosu.RowException.
Information about an Exception related to a row in a table.

RowException() - Constructor for class org.iscmem.cosu.RowException

run(DoubleTable, ByReferenceBoolean) - Method in interface org.iscmem.cosu.Operation
Perform the Operation.

S

SelfDescribingOperation - interface org.iscmem.cosu.SelfDescribingOperation.
   An interface that allows Operations to describe themselves.

setBooleanAt(int, int, boolean) - Method in interface org.iscmem.cosu.SimpleTable
   Set the value in the given cell.
setBooleans(int, boolean[]) - Method in interface org.iscmem.cosu.SimpleTable
   Set the values in the given row.

setDoubleAt(int, int, double) - Method in interface org.iscmem.cosu.DoubleTable
   Put a value in the table.

setDoubles(int, double[]) - Method in interface org.iscmem.cosu.DoubleTable
   Fill a row in the table.

setIntAt(int, int, int) - Method in interface org.iscmem.cosu.SimpleTable
   Set the value in the given cell.

setInts(int, int[]) - Method in interface org.iscmem.cosu.SimpleTable
   Set the values in the given row.

setObjectAt(int, int, Object) - Method in interface org.iscmem.cosu.ComplexTable
   Set the value in the given cell.

setObjects(int, Object[]) - Method in interface org.iscmem.cosu.ComplexTable
   Set the values in the given row.

setRowName(int, String) - Method in interface org.iscmem.cosu.DoubleTable
   Set the name for the given row.

setStringAt(int, int, String) - Method in interface org.iscmem.cosu.SimpleTable
   Set the value in the given cell.

setStrings(int, String[]) - Method in interface org.iscmem.cosu.SimpleTable
   Set the values in the given row.

setUp() - Method in interface org.iscmem.cosu.Operation
   Before calling run or restart the first time, setUp should be called.

SimpleTable - interface org.iscmem.cosu.SimpleTable.
   This type of table can store floating point numbers (doubles), booleans, integers, and
   strings.

SimpleTransformation - interface org.iscmem.cosu.SimpleTransformation.
An interface for a Transformation that is applied to a SimpleTable and produces a SimpleTable.

stop() - Method in interface org.iscmem.cosu.Operation
Request that the operation stop prematurely.

stopExecution() - Method in interface org.iscmem.cosu.Executer
Stop Execution to the extent feasible.

STRING - Static variable in interface org.iscmem.cosu.SimpleTable

supportsParallelRuns() - Method in interface org.iscmem.cosu.Operation
Indicate if multiple instances of the Operation can be performed in parallel.

T

TableException - exception org.iscmem.cosu.TableException.
An Exception thrown when one of the semantics of Operations on tables has been violated (e.g., close a table that has not been opened, access a table that has not been opened, access a value that has not been set).

TableException() - Constructor for class org.iscmem.cosu.TableException
Creates a new instance of TableException without detail message.

TableException(String) - Constructor for class org.iscmem.cosu.TableException
Constructs an instance of TableException with the specified detail message.

transform(ComplexTable) - Method in interface org.iscmem.cosu.ComplexTransformation
Apply the ComplexTransformation and return the result.

transform(DoubleTable) - Method in interface org.iscmem.cosu.DoubleTransformation
Apply the DoubleTransformation and return the result.

transform(SimpleTable) - Method in interface org.iscmem.cosu.SimpleTransformation
Apply the SimpleTransformation and return the result.

type - Variable in class org.iscmem.cosu.Column
Information about the type of information stored in the column.

V

value - Variable in class org.iscmem.cosu.ByReferenceBoolean

W

waitForDataAvailable(int, int) - Method in interface org.iscmem.cosu.DoubleTable
    Wait for a value to become available in the given cell.

waitForDone() - Method in interface org.iscmem.cosu.Executer
    Return when all queued Operations have completed.

B C D E F G I N O Q R S T V W
Package Class **Tree Deprecated** **Index** **Help**
PREV NEXT   FRAMES NO FRAMES   All Classes

Appendix B: Selected Workshop Bibliography

1. Andres, T.H., 1997, "Sampling Methods and Sensitivity Analysis for Large Parameter Sets," *Journal of Statist. Comput. Simul.*, Vol. 57, p.77–110.

2. Andres, T.H., 1998, "SAMPLE2: Software to Generate Experimental Design for Large Sensitivity Analysis Experiments." *Proc. of the Second International Symposium on Sensitivity Analysis of Model Output*, Venice, Ca'Dolfin 1998-April 19–22.

3. Andres, T.H. 1993. " Using Iterated Fractional Factorial Design to Screen Parameters in Sensitivity Analysis of a Probabilistic Risk Assessment Model," *Proc. Joint International Conference on Mathematical Methods and Supercomputing in Nuclear Applications*, Karlsruhe, Germany, 1993 April 19–23, Vol. 2 p. 328–37.

4. Beck, M.B., 1987, "Water Quality Modeling: A Review of the Analysis of Uncertainty," *Water Resources Research*, 23(8), pp 1393–1442.

5. Beck, M.B., and J. Chen, 2000, "Assuring the Quality of Models Designed for Predictive Tasks," in Sensitivity Analysis (A. Saltelli, K. Chan, and E.M. Scott, eds), Wiley, Chichester, pp 401–420.

6. Beck, M.B., ed., 2002, "Environmental Foresight and Models: A Manifesto," Elsevier, Oxford.

7. Beven, K.J., and J. Freer, 2001, Equifinality, data assimilation, and uncertainty estimation in mechanistic modelling of complex environmental systems, *J. Hydrology*, 249, 11–29.

8. Beven, K.J., 2002, Towards a coherent philosophy for environmental modelling, *Proc. Roy. Soc. Lond.* A, 458, 2465–2484.

9. Beven, K.J., and J. Feyen, 2002, The future of distributed hydrological modelling, *Hydrol. Process.*, 16(2), 169–172.

10. Beven, K.J., 2002, Towards an alternative blueprint for a physically based digitally simulated hydrologic response modelling system, *Hydrol. Process.*, 16(2), 189–206.

11. Borsuk, M.E., C.A. Stow, and K.H. Reckhow, 2003, "An integrated approach to TMDL development for the Neuse River Estuary using a Bayesian probability network model (Neu-BERN)," *Journal Water Resources Planning and Management*. In press. (July issue).

12. Borsuk, M.E., C.A. Stow, and K.H. Reckhow, 2002, Predicting the frequency of water quality standard violations: A probabilistic approach for TMDL development. *Environmental Science and Technology*. 36:2109–2115.

13. Cullen, A.C., and H.C. Frey, 1999, *The Use of Probabilistic Techniques in Exposure Assessment: A Handbook for Dealing with Variability and Uncertainty in Models and Inputs.* Plenum: New York. 335 pages.

14. Frey, H.C., and D.S. Rhodes, 1996, "Characterizing, Simulating, and Analyzing Variability and Uncertainty: An Illustration of Methods Using an Air Toxics Emissions Example," *Human and Ecological Risk Assessment: an International Journal*, 2(4):762–797 (December).

15. Frey, H.C., and D.S. Rhodes, 1998, "Characterization and Simulation of Uncertain Frequency Distributions: Effects of Distribution Choice, Variability, Uncertainty, and Parameter Dependence," *Human and Ecological Risk Assessment: an International Journal*, 4(2):423–468 (April).

16. Frey, H.C., and D.E. Burmaster, 1999, "Methods for Characterizing Variability and Uncertainty: Comparison of Bootstrap Simulation and Likelihood-Based Approaches," *Risk Analysis*, 19(1):109–130 (February).

17. Frey, H.C., and S.R. Patil, 2002, "Identification and Review of Sensitivity Analysis Methods," *Risk Analysis*, 22(3):553–578 (June).

18. Helton, J.C., J.W. Garner, R.D. McCurley, and D.K. Rudeen, 1991, *Sensitivity Analysis Techniques and Results for Performance Assessment at the Waste Isolation Pilot Plant*, SAND90-7103, Albuquerque, New Mexico.

19. Helton, J.C., 1993, Uncertainty and Sensitivity Analysis Techniques for Use in Performance Assessment for Radioactive Waste Disposal, *Reliability Engineering and System Safety* 42 (2-3): 327–367.

20. Hill, M.C., 1998, "Methods and guidelines for effective model calibration," U.S Geological Survey Water-Resources Investigations Report 98-4005, 90p. http://pubs.water.usgs.gov/wri984005/.

21. Leavesley, G.H., L.E. Hay, R.J. Viger, and S.L. Markstrom, 2003, "Use of *a priori* parameter-estimation methods to constrain calibration of distributed-parameter models, in *Calibration of watershed models: American Geophysical Union, Water Science and Application* 6, (Q. Duan, H.V. Gupta, S. Sorooshian, A.N. Rousseau, and R. Turcotte, eds.) p. 255–266.

22. Lu, Y., and S. Mohanty, 2001, "Sensitivity Analysis of a Complex, Proposed Geologic Waste Disposal System Using the Fourier Amplitude Sensitivity Test Method." *Reliability Engineering and System Safety*. 72 (3) p. 275–291.

23. McKay, Michael D., Richard J. Beckman, Leslie M. Moore, and Richard R. Picard, "An Alternative View of Sensitivity in the Analysis of Computer Codes," in *Proceedings Section on Physical and Engineering Sciences Section of the American Statistical Association, Boston, August 9-13, 1992*, LAUR 92–1585, Los Alamos National Laboratory, Los Alamos, New Mexico.

24. McKay, Michael D., 1996, "Application of Variance-Based Methods to NUREG-1150 Uncertainty Analysis" (prepared for U.S. NRC), LAUR 96–145, Los Alamos National Laboratory, Los Alamos, New Mexico, (June 28).

25. Mehl, S.W., and M.C. Hill, 2001, "A comparison of solute-transport solution techniques and their effect on sensitivity analysis and inverse modeling results," *Ground Water*, 39(2): 300–307.

26. Meyer, P.D., M.L. Rockhold, and G.W. Gee, 1997, "Uncertainty Analyses of Infiltration and Subsurface Flow and Transport for SDMP Sites," NUREG/CR-6565, U.S. Nuclear Regulatory Commission, Washington, DC. (http://nrc-hydro-uncert.pnl.gov/)

27. Meyer, P.D. and G.W. Gee, 1999, "Information on Hydrologic Conceptual Models, Parameters, Uncertainty Analysis, and Data Sources for Dose Assessments at Decommissioning Sites," NUREG/CR-6656, U.S. Nuclear Regulatory Commission, Washington, DC. (http://nrc-hydro-uncert.pnl.gov/).

28. Meyer, P.D., and R.W. Taira, 2001, "Hydrologic Uncertainty Assessment for Decommissioning Sites: Hypothetical Test Case Applications," NUREG/CR-6695, U.S. Nuclear Regulatory Commission, Washington, DC. (http://nrc-hydro-uncert.pnl.gov/)

29. Minsker, Barbara, ed., 2003, "Long-Term Ground-Water Monitoring: The State-of-the-Art" American Society of Civil Engineers, stock number 40678, 116p, http://www.pubs.asce.org/BOOKdisplay.cgi?9991614.

30. Mohanty. S., Y. Lu, and J.M. Menchaca, "Preliminary Analysis of Morris Method for Identifying Influential Parameters," *American Nuclear Society Transactions*, Vol. 81, p. 55–56.

31. Mohanty, S., and Y-T. (Justin) Wu, 2001, "CDF Sensitivity Analysis Technique for Ranking Influential Parameters in the Performance Assessment of the Proposed High-Level Waste Repository at Yucca Mountain, Nevada, USA," *Reliability Engineering and System Safety*, **73**, 2, p. 167.

32. Mohanty, S., and Y-T. (Justin) Wu, 2002, *"Mean-based Sensitivity or Uncertainty Importance Measures for Identifying Influential Parameters,"* *Probabilistic Safety Assessment and Management (PSAM6)*, Bonano, E.J., A.L. Camp, M.J. Majors, R.A. Thompson (eds.), Vol. 1, p. 1079–1085, Elsevier: New York, New York, USA.

33. Mohanty, S., and R. Codell, 2002, "Sensitivity Analysis Methods for Identifying Influential Parameters in a Problem with a Large Number of Random Variables," *Risk Analysis III* (C. Brebbia, ed.) WIT Press, Boston, USA: 363–374.

34. Morris, M.D., 1991, "Factorial sampling plans for preliminary computational experiments," *Technometrics* 33 (2): 161–174.

35. Neuman, S.P. and P.J. Wierenga, 2003, *"A Comprehensive Strategy of Hydrogeologic Modeling and Uncertainty Analysis for Nuclear Facilities and Sites,"* NUREG/CR-6805, U.S. Nuclear Regulatory Commission, Washington, DC.

36. Osidele, O.O., W. Zeng, and M.B. Beck, 2003, "Coping with Uncertainty: A Case Study in Sediment Transport and Nutrient Load Analysis," *Journal of Water Resources Planning and Management*, 129(4): 1–11.

37. Poeter, E.P., and M.C. Hill, 1998, *Documentation of UCODE, A Computer Code for Universal Inverse Modeling*, U.S. Geological Survey Water-Resources Investigations Report 98-4080, 116 pp., U.S. Geological Survey, Denver, Colorado.

38. Reckhow, K.H, 1994, "Water Quality Simulation Modeling and Uncertainty Analysis for Risk Assessment and Decision Making," *Ecological Modeling* 72:1–20.

39. Saltelli, A., and S. Tarantola, 2002, "On the relative importance of input factors in mathematical models: safety assessment for nuclear waste disposal," *Journal of American Statistical Association*, 97 (459), 702–709.

40. Saltelli, A., K. Chan, and E.M. Scott (eds.), 2000, "Sensitivity Analysis", *Wiley Series in Probability and Statistics*, Chichester, New York: Wiley.

41. Saltelli, A., 2002, "Making best use of model valuations to compute sensitivity indices." *Computer Physics Communications*, 145, 280–297.

42. Tarantola S., and A. Saltelli, 2003, "SAMO 2001: Methodological advances and innovative applications of sensitivity analysis," *Reliab. Engng. Syst. Safety*, 79 (2), 121–122.

43. Tiedeman, C.R., M.C. Hill, F.A. D'Agnese, and C.C. Faunt, 2003, "Methods for using groundwater model predictions to guide hydrogeologic data collection, with application to the Death Valley regional ground-water flow system," *Water Resources Research*, 39(1):10.1029/2 001WR001255.

Appendix C: Selected Web Site Links

1. MOU Public Web site: http://ISCMEM.Org

2. PNNL Web site for Uncertainty Research: http://nrc-hydro-uncert.pnl.gov/

3. Andrea Saltelli, Applied Statistics Web site: http://www.jrc.cec.eu.int/uasa and forum for sensitivity analysis: http://sensitivity-analysis.jrc.cec.eu.int/

4. NUREG/CR-6565, "Uncertainty Analyses of Infiltration and Subsurface Flow and Transport for SDMP Sites," at http://www.nrc.gov/reading-rm/doc-collections/nuregs/contract/cr6565/

5. NUREG/CR-6767, "Evaluation of Hydrologic Uncertainty Assessments for Decommissioning Sites Using Complex and Simplified Models," at http://www.nrc.gov/reading-rm/doc-collections/nuregs/contract/cr6767/

6. NUREG/CR-6805, "A Comprehensive Strategy of Hydrogeologic Modeling and Uncertainty Analysis for Nuclear Facilities and Sites," at http://www.nrc.gov/reading-rm/doc-collections/nuregs/contract/cr6805/

7. NUSAP (Numeral, Unit, Spread, Assessment, Pedigree) - The Management of Uncertainty and Quality in Quantitative Information Web site: http://www.nusap.net

Appendix D: List of Attendees by Organization

Organization	Last Name	First Name
Applied Biomathematics	Ferson	Scott
Boise State University	Clemo	Thomas M.
Center for Nuclear Waste Regulatory Analyses	LaPlante	Patrick A.
Center for Nuclear Waste Regulatory Analyses	Mohanty	Sitakanta
Center for Nuclear Waste Regulatory Analyses	Pensado	Osvaldo, DR.
Colorado School of Mines	Poeter	Eileen, P
Copernicus Institute for Sustainable Development and Innovation, Utrecht University	Van der Sluijs	Jeroen P.
Duke University	Reckhow	Kenneth H.
European Commission - Joint Research Centre	Saltelli	Andrea
Geological Survey of Denmark and Greenland	Refsgaard	Jens Christian
Idaho National Engineering and Environmental Laboratories	Palmer	Carl
Idaho National Engineering and Environmental Laboratories	Schafer	Annette L.
Idaho National Engineering and Environmental Laboratories	Shook	Michael G.
Lancaster University	Beven	Keith J.
National Oceanic and Atmospheric Administration	Duan	Qingyun
National Oceanic and Atmospheric Administration	Hicks	Bruce B.
National Oceanic and Atmospheric Administration	Whitall	David R.
Neptune and Company, Inc.	Black	Paul K.
Neptune and Company, Inc.	Tauxe	John
North Carolina State University - Department of Civil, Construction, and Environmental Engineering	Frey	H. Christopher
Pacific Northwest National Laboratory	Castleton	Karl J.
Pacific Northwest National Laboratory	Eslinger	Paul W.
Pacific Northwest National Laboratory	Meyer	Philip D.
Sandia National Laboratories	Criscenti	Louise J.
Sandia National Laboratories	Helton	Jon
Sandia National Laboratories	Roberts	Randall
U.S. Army Corps of Engineers	Edris	Earl V.
U.S. Army Corps of Engineers	Skahill	Brian E.
U.S. Army Corps of Engineers	Zakikhani	Mansour
U.S. Army Corps of Engineers, Engineering Research and Development Center	Bunch	Barry W.

Organization	Last Name	First Name
U.S. Army Corps of Engineers, Engineering Research and Development Center	Dortch	Mark S.
U.S. Army Corps of Engineers, Engineering Research and Development Center Coastal and Hydraulics Laboratories	Wallace	Robert M.
U.S. Army Corps of Engineers, Engineering Research and Development Center Waterways Experiment Center	Bridges	Todd S.
U.S. Department of Agriculture	Ahuja	Laj R.
U.S. Department of Agriculture	Seyfried	Mark S.
U.S. Department of Agriculture	Weltz	Mark A.
U.S. Department of Agriculture - Agriculture Research Service	Ascough	James C., II
U.S. Department of Agriculture - Agriculture Research Service	Flanagan	Dennis C.
U.S. Department of Agriculture - Agriculture Research Service	Ma	Liwang
U.S. Department of Agriculture - Agriculture Research Service	Savabi	Reza M.
U.S. Department of Agriculture - Agriculture Research Service, Environmental Microbial Safety Laboratory	Pachepsky	Yakov
U.S. Department of Energy	Moore	Beth A.
U.S. Department of Energy	van Luik	Abraham
U.S. Department of Energy - Science Application International Corporation	Kahn	Alauddin
U.S. Environmental Protection Agency	Fite	Edward C.
U.S. Environmental Protection Agency	Griffin	Susan
U.S. Environmental Protection Agency	Kroner	Stephen M.
U.S. Environmental Protection Agency	Langstaff	John E.
U.S. Environmental Protection Agency	Laniak	Gerry
U.S. Environmental Protection Agency	Parmar	Rajbir S.
U.S. Environmental Protection Agency	Pascual	Pasky A.
U.S. Environmental Protection Agency	Shenk	Gary W.
U.S. Environmental Protection Agency	Stiber	Neil
U.S. Environmental Protection Agency	Wolfe	Kurt L.
U.S. Environmental Protection Agency	Young	Dirk F.
U.S. Environmental Protection Agency - Office of Science Policy	Sunderland	Elsie M.
U.S. Environmental Protection Agency - Office of Solid Waste and Emergency Response	Chang	Steven

Organization	Last Name	First Name
U.S. Geological Survey	Bales	Jerad D.
U.S. Geological Survey	Banta	Edward R.
U.S. Geological Survey	Gutierrez-Magness	Angelica
U.S. Geological Survey	Leavesley	George H.
U.S. Geological Survey	Markstrom	Steven L.
U.S. Geological Survey	Mason	Robert R.
U.S. Geological Survey	Pollock	David W.
U.S. Geological Survey	Tiedeman	Clare R.
U.S. Geological Survey- Water Resources Discipline	Holtschlag	David J.
U.S. Geological Survey- Water Resources Discipline	Raffensperger	Jeff P.
U.S. Nuclear Regulatory Commission	Abu-Eid	Boby
U.S. Nuclear Regulatory Commission	Cady	Ralph E.
U.S. Nuclear Regulatory Commission	Codell	Richard B.
U.S. Nuclear Regulatory Commission	Damon	Dennis
U.S. Nuclear Regulatory Commission	Esh	David W.
U.S. Nuclear Regulatory Commission	McCartin	Tim
U.S. Nuclear Regulatory Commission	Nicholson	Thomas
U.S. Nuclear Regulatory Commission	Peckenpaugh	Jon M.
U.S. Nuclear Regulatory Commission	Salomon	Arthur D.
U.S. Nuclear Regulatory Commission	Strosnider	Jack R.
U.S. Nuclear Regulatory Commission	Thaggard	Mark
University of Arizona	Neuman	Shlomo P.
University of Georgia	Beck	Bruce M.
University of Georgia	Osidele	Olufemi O.
University of Illinois	Valocchi	Albert
University of Manitoba	Andres	Terry H.
University of Queensland and S.S. Papadopulos & Associates	Tonkin	Matthew J.
University of Riverside Department of Environmental Sciences	van Griensven	Ann
Utah State University - Department of Civil and Environmental Engineering	Bastidas	Luis A.
Watermark Numerical Computing	Doherty	John
Westinghouse Savannah River Company	Flach	Gregory P.

Appendix E: Agenda

International Workshop on Uncertainty, Sensitivity, and Parameter Estimation for Multimedia Environmental Modeling

Dates: August 19–21, 2003

Location: U.S. Nuclear Regulatory Commission (NRC) Headquarters Auditorium, 11545 Rockville Pike, Rockville, Maryland, USA

Sponsorship: The Federal Working Group on Uncertainty and Parameter Estimation[1] under the Federal Interagency Steering Committee on Multimedia Environmental Modeling (ISCMEM)

Technical Topics: **Uncertainty Analysis**, **Sensitivity Analysis** and **Parameter Estimation** Related to **Multimedia Environmental Modeling**

Workshop Objectives: Facilitate communication among U.S. Federal agencies conducting research on the workshop themes, obtain up-to-date information from invited technical experts, and actively discuss opportunities and new approaches for parameter estimation, and sensitivity and uncertainty analyses related to multimedia environmental modeling.

Attendance: All MOU[1] participating Federal agencies, invited speakers, and sponsored technical specialists.

Registration: No registration fee, but prior registration is required. All registrants must be sponsored by one of the eight MOU parties. Due to a limited number of registration spaces, registrants are encouraged to attend all 3 days of the workshop. To access through NRC security to attend the workshop, all attendees must have photo IDs for U.S. citizens, and passports for non-U.S. citizens. Please email address and contact information to workshop_uncertainty@nrc.gov.

Documentation: Abstracts along with viewgraphs or PowerPoint presentations are requested 2 weeks prior to the workshop.

Proceedings: Summary of meeting discussions and presentations, as extended abstracts with supporting technical references and Web sites, and proposal for an international conference to be held in 2004 will be posted on the MOU public Web site: http://ISCMEM.Org.

---

[1] Detailed information on membership, activities, and technical background for the Memorandum of Understanding (MOU), and its Federal working groups (FWGs) can be found on the public Web site: http://ISCMEM.Org.

August 19

9:00 a.m.	Welcome and Opening Remarks Jack Strosnider, Deputy Director, Office of Nuclear Regulatory Research, U.S. Nuclear Regulatory Commission (NRC)
9:15	Introduction of the Workshop Objectives, Technical Themes, and Goals George Leavesley, U.S. Geological Survey (USGS) and Co-Chair, Federal Working Group on Uncertainty and Parameter Estimation (FWG)
9:30	Federal Agency Overviews of Parameter Estimation, Sensitivity and Uncertainty Approaches [focus on agency's motivation, activities, capabilities, and research related to the workshop themes (15 minutes each)] U.S. Nuclear Regulatory Commission ....Tom Nicholson, NRC/RES U.S. Environmental Protection Agency ....Justin Babendreier, EPA U.S. Geological Survey .... George Leavesley, USGS
10:15	BREAK
10:35	Federal Agency Overviews (continue) National Oceanic & Atmospheric Administration .... Bruce Hicks, NOAA U.S. Department of Energy .... Beth Moore, DOE USDA/Agricultural Research Service.. Mark Weltz, ARS U.S. Army Corps of Engineers .... Earl Edris, USACOE
11:35	LUNCH

Session Theme: Parameter Estimation Approaches, Applications, and Lessons Learned — Identification of Research Needs

Session Facilitator: Earl Edris, USACOE   Session Rapporteur: Phil Meyer, PNNL

12:40 p.m.	Unsaturated Zone Parameter Estimation Using HYDRUS and Rosetta Codes Rien van Genuchten and Jirka Simunek, ARS
1:05	Parameter Estimation and Predictive Uncertainty Analysis for Ground and Surface Water Models using PEST John Doherty, Watermark Numerical Computing, Inc., Australia
1:35	A Priori Parameter Estimation: Issues and Uncertainties George Leavesley, USGS
2:00	Multi-Objective Approaches for Parameter Estimation and Uncertainty Luis Bastidas, Utah State University
2:30	BREAK
2:50	Using Sensitivity Analysis in Model Calibration Efforts Claire R. Tiedeman and Mary C. Hill, USGS
3:15	Jupiter Project—Merging Inverse Problem Formulation Technologies Mary Hill, Eileen Poeter*, Colorado School of Mines, J. Doherty and Ned Banta

3:40 p.m.	Simulated Contaminant Plume Migration: The Effects of Geochemical Parameter Uncertainty Louise J. Criscenti, Mehdi Eliassi, Randall T. Cygan, and Malcolm D. Siegel, Sandia National Laboratory
4:05	Impact of Sensitive Parameter Uncertainties on Dose Impact Analysis for Decommissioning Sites Boby Abu-Eid and Mark Thaggard, NRC
4:25	Discussion of Parameter Estimation Approaches and Applications (Rapporteur & Facilitator)
5:30	ADJOURN

August 20

8:30 a.m.	Review Agenda and Announcements. T. Nicholson, USNRC and FWG Co-Chair

Session Theme: Sensitivity Analysis Approaches, Applications, and Lessons Learned — Identification of Research Needs

Session Facilitator: Tom Nicholson, NRC   Session Rapporteur: Sitakanta Mohanty, CNWRA

8:45 a.m.	Global Sensitivity Analysis: Novel Settings and Methods Andrea Saltelli, European Commission Joint Research Center, Italy
9:25	Sampling-Based Approaches to Uncertainty and Sensitivity Analysis Jon Helton, Arizona State University
9:55	Uncertainty and Sensitivity Analysis for Environmental and Risk Assessment Models Christopher Frey, North Carolina State University
10:25	BREAK
10:45	Practical Strategies for Sensitivity Analysis Given Models with Large Parameter Sets Terry Andres, University of Manitoba, Canada
11:15	An Integrated Regional Sensitivity Analysis and Tree-Structured Density Estimation Methodology Femi Osidele and Bruce Beck, University of Georgia
11:45	Uncertainty and Sensitivity Analyses in the Context of Determining Risk Significance Sitakanta Mohanty, CNWRA
12:10 p.m.	Discussion of Sensitivity Approaches and Applications with Emphasis on Relationship to Parameter Estimation and Uncertainty (Rapporteur & Facilitator)
12:30	LUNCH
1:20	Discussion of Sensitivity Approaches and Applications with Emphasis on Relationship to Parameter Estimation and Uncertainty (Continued)

Session Theme: Uncertainty Analysis Approaches, Applications, and Lessons Learned — Identification of Research Needs

Session Facilitator: Rien van Genuchten, ARS   *Session Rapporteur: Sitakanta Mohanty, CNWRA*

1:40 p.m.	Uncertainty: Foresight, Evaluation, and System Identification Bruce Beck, University of Georgia
2:10	Uncertainty in Catchment Modeling: A Manifesto for Equifinality Keith Beven, University of Lancaster, United Kingdom
2:40	Model Abstraction Techniques Related to Parameter Estimation and Uncertainty Yakov Pachepsky, ARS
3:05	BREAK
3:25	Toward a Synthesis of Qualitative and Quantitative Uncertainty Assessment: Applications of the Numeral, Unit, Spread, Assessment, Pedigree (NUSAP) System Jeroen van der Sluijs, Copernicus Institute for Sustainable Development and Innovation, Utrecht University, The Netherlands
3:55	Hydrogeologic Conceptual Model and Parameter Uncertainty. Shlomo Neuman,University of Arizona
4:25	Development of a Unified Uncertainty Methodology Phil Meyer, Pacific Northwest National Laboratory
4:50	Discussion of Uncertainty Approaches and Applications (Rapporteur & Facilitator)
5:30	ADJOURN

August 21

8:15 a.m.	Review Agenda and Announcements, G. Leavesley, USGS and FWG Co-Chair

Session Theme: Parameter Estimation, Sensitivity and Uncertainty Approaches — Applications and Lessons Learned

Session Facilitator: George Leavesley, USGS   *Session Rapporteur: Bruce Hicks, NOAA*

8:30 a.m.	Probabilistic Risk Assessment for Total Maximum Daily Surface-Water Loads Ken Reckhow, Duke University
9:00	A Stochastic Risk Model for the Hanford Nuclear Site Paul W. Eslinger, Pacific Northwest National Laboratory
9:25	National-Scale Multimedia Risk Assessment for Hazardous Waste Disposal Justin Babendreier, EPA
9:50	BREAK
10:10	Ground-Water Parameter Estimation and Uncertainty Applications Earl Edris, USACOE

10:35	Use of Fractional Factorial Design for Sensitivity Studies Richard Codell, NRC
11:00	ISCORS Parameter-Source Catalog Anthony B. Wolbarst, EPA
11:20	Roundtable Discussion by Session Facilitator and Rapporteurs Focusing on List of Salient Points Identified During the Workshop and Suggestions on Future Directions for Parameter Estimation, Sensitivity and Uncertainty Research
12:10 p.m.	LUNCH

Session Theme: Toward Development of a Common Software Application Programming Interface (API) for Uncertainty, Sensitivity, and Parameter Estimation Methods and Tools

Afternoon Working Session (All Workshop Participants Are Strongly Encouraged to Attend)

Session Facilitator: George Leavesley, USGS   Session Rapporteur: Justin Babendreier, EPA

1:00 p.m.	Introduction of the Uncertainty Analysis, Sensivitivy Analysis and Parameter Estimation (UA/SA/PE) API Session Objectives and Technical Goal George Leavesley, USGS and FWG Co-Chair
1:10	The Related Role of Environmental Modeling Frameworks Gerry Laniak, EPA and Co-Chair, Federal Working Group on Frameworks and Technology
1:35	Conceptual Structure for a Common UA/SA/PE API Karl Castleton, PNNL; Steve Fine, EPA
2:00	Themes for Audience Discussion Moderator, George Leavesley, USGS and FWG Co-Chair

1. Why is the UA/SA/PE API important to non-programmers?

2. How important is nesting of operations?

3. Are tables sufficient for data exchange between UA/SA/PE components?

4. Where are the logical connections between UA/SA/PE components (i.e., where are tables produced and consumed)?

5. How should UA/SA/PE components be run?

Open Audience Discussion: Building Consensus on UA/SA/PE API Structure (Technologist-to-Scientist Discussions)

2:40	BREAK
3:00	Open Discussion (continued) Facilitator: George Leavesley, USGS and Rapporteur: Justin Babendreier, EPA
3:50	Closing Remarks Mark Dortch, USACOE and Chair, Federal Interagency Steering Committee
4:00	ADJOURN

Program Format:

- Each presenter is encouraged to provide an extended abstract (200 words minimum up to 6 pages maximum) along with a list of keywords, Web site links, and references for distribution prior to the workshop.

- The program is organized into four thematic sessions on parameter estimation, sensitivity, uncertainty, and applications; each session highlights invited talks (30 minutes) by selected experts and contributed papers (20–25 minutes) on applications that focus on the technical theme;

- An extended discussion period will be provided at the end of each thematic session.

- Session rapporteurs will list methods, approaches, and applications identified, with emphasis on practical strategies for each theme.

- Attendees will have an opportunity to provide written questions and suggestions to the session rapporteurs during breaks before the discussion periods;

- A roundtable discussion by the session rapporteurs and facilitators will summarize the workshop's overall technical ideas and themes for consideration in proposing an international conference.

- Final working session for "Technologist-to-Scientist" discussions will focus on development of a common software application programming interface (API) for uncertainty analysis, sensitivity analysis, and parameter estimation methods and tools.

NRC FORM 335
(9-2004)
NRCMD 3.7

U.S. NUCLEAR REGULATORY COMMISSION

BIBLIOGRAPHIC DATA SHEET

(See instructions on the reverse)

1. REPORT NUMBER
(Assigned by NRC, Add Vol., Supp., Rev., and Addendum Numbers, if any.)

NUREG/CP-0187
ERDC SR-04-2
EPA/600/R-04/117

2. TITLE AND SUBTITLE

Proceedings of the International Workshop on Uncertainty, Sensitivity, and Parameter Estimation for Multimedia Environmental Modeling

Held August 19-21, 2003 at the U.S. Nuclear Regulatory Commission Headquarters, Rockville, Maryland, USA

3. DATE REPORT PUBLISHED

MONTH	YEAR
October	2004

4. FIN OR GRANT NUMBER

5. AUTHOR(S)

T. J. Nicholson (NRC), J. Babendreier (EPA), P. Meyer (PNNL), S. Mohanty (CNWRA), B. Hicks (NOAA), and G. Leavesley (USGS)

6. TYPE OF REPORT

Technical Proceedings

7. PERIOD COVERED *(Inclusive Dates)*

August 19-21, 2003

8. PERFORMING ORGANIZATION - NAME AND ADDRESS *(If NRC, provide Division, Office or Region, U.S. Nuclear Regulatory Commission, and mailing address; if contractor, provide name and mailing address.)*

Federal Working Group on Uncertainty and Parameter Estimation

Interagency Steering Committee on Multimedia Environmental Models

c/o T. Nicholson, DSARE/ RES/ U.S. NRC at Mail Stop T-9C34, 11545 Rockville Pike, Rockville, MD 20852-2738

9. SPONSORING ORGANIZATION - NAME AND ADDRESS *(If NRC, type "Same as above"; if contractor, provide NRC Division, Office or Region, U.S. Nuclear Regulatory Commission, and mailing address.)*

Division of Systems Analysis and Regulatory Effectiveness

Office of Nuclear Regulatory Research

U.S. Nuclear Regulatory Commission

Washington, DC 20555-0001

10. SUPPLEMENTARY NOTES

ISCMEM Publication to be placed on ISCMEM Web site: http://www.ISCMEM.Org

11. ABSTRACT *(200 words or less)*

An International Workshop on Uncertainty, Sensitivity, and Parameter Estimation for Multimedia Environmental Modeling was held August 19-21, 2003, at the U.S. Nuclear Regulatory Commission Headquarters in Rockville, Maryland, USA. The workshop was organized and convened by the Federal Working Group on Uncertainty and Parameter Estimation, and sponsored by the Federal Interagency Steering Committee on Multimedia Environmental Models (ISCMEM). The workshop themes were parameter estimation, sensitivity analysis, and uncertainty analysis relevant to environmental modeling. The workshop objectives were to facilitate communication among U.S. Federal agencies conducting research on the workshop themes; obtain up-to-date information from invited technical experts; actively discuss the state-of-the-science in the workshop themes; and identify opportunities for pursuing new approaches. These objectives were met through the workshop presentations and discussions. The invited presenters focused on methods to identify, evaluate, and compare both existing and newly developed strategies and tools for parameter estimation, sensitivity and uncertainty analyses. Discussions explored how these strategies and tools could be used to better understand and characterize the sources of uncertainty in environmental modeling, and approaches to quantify them through comparative analysis of model simulations and monitoring. The presentations and discussions also focused on various approaches and applications of these strategies and tools, and specific lessons learned and research needs. In addition, the Memorandum of Understanding working group members and cooperators presented information and guidance for use in developing a common software application programming interface for methods and tools used in parameter estimation, sensitivity analysis, and uncertainty analysis.

12. KEY WORDS/DESCRIPTORS *(List words or phrases that will assist researchers in locating the report.)*

application programming interface
conceptual model uncertainty
environmental modeling
model calibration
model evaluation
optimization
parameter estimation
predictive uncertainty analysis
sensitivity analysis
uncertainty assessment

13. AVAILABILITY STATEMENT

unlimited

14. SECURITY CLASSIFICATION

(This Page)

unclassified

(This Report)

unclassified

15. NUMBER OF PAGES

16. PRICE

PRINTED ON RECYCLED PAPER

Federal Recycling Program